Advances in ... y of seeds
Part 6

Advances in research and technology of seeds

Part 6

edited for the International Seed Testing Association

by J.R. Thomson

pudoc

Centre for Agricultural Publishing and Documentation

Wageningen - 1981

ISBN 90-220-0786-3

List of contributors

Dr P.A. Thompson
Royal Botanic Garden Kew
Ardingley
HAYWARDS HEATH
RH17 6TN
UK

Dr Daphne J. Osborne
Weed Research Organisation
Begbrooke Hill
OXFORD
OX5 1PF
UK

Dr J.C. Slaughter
Department of Brewing & Biological Sciences
Heriot-Watt University
Chambers Street
EDINBURGH
EH1 1HX
UK

Prof. A. Lovato
Instituto di Agronomia Generale e Coltivazioni Erbacee
40126 BOLOGNA
Via Filippo Re, 6-8
Italy

Contents

Introduction

This issue of ARTS is mainly taken up with reviews of the continuing flow of papers in the literature on the biochemistry and physiology of seed development and germination. Looking at germination from a different angle we welcome a contribution from Dr Thompson of Kew Gardens with reference to ecology. This follows on our review in Part 2 which elicited encouraging comments.

J.R. Thomson
The Edinburgh School of Agriculture

Ecological aspects of seed germination

P.A. Thompson

It is widely accepted that germination responses are strongly selected for under natural conditions, with the result that different populations of a species from different parts of its range possess responses to temperature and other environmental factors which result from adaptations to locally significant selective pressures. In fact this viewpoint is largely based on instinct. There is relatively little data to support it and virtually none which unequivocally illustrates processes of natural selection occurring in the germination processes of wild species. The review which follows attempts to assess recent evidence bearing on the natural germination responses of seeds of wild species, and to consider these responses in relation to variations in seed morphology, reproductive capacity and the stresses which result from predator activity, and climatic or edaphic conditions. The main aims of the review are an examination of the nature of the adaptive responses of seeds and, in particular, of the factors which might lead to relative stability, or plasticity, in rates of change of the processes which are involved in the control of germination.

Stability of germination responses

Direct evidence for rapid changes in germination behaviour in response to selection is relatively scarce. Morley (1958) found that the hard seed coat character, which imposes dormancy in *Trifolium subterraneum*, could be modified after a few generations in cultivation so as to produce populations in which high proportions of seeds possessed water permeable seed coates, with few restrictions on germination. Whitcombe & Whittington (1972) have also demonstrated progressive reductions in an undefined dormancy in *Sinapis alba* seeds from one generation to another, under conditions favouring selection for spontaneous germination. The genetic control of germination processes has been reviewed as a topic by Whittington (1973) who concluded that germination characteristics are at least partially under genetic control, and that the pattern of behaviour in cultivated or wild species is likely to have resulted from selection. This tentative phrasing epitomises the quality of evidence available. Thompson & Fox (1976) have suggested that the temperature-mediated germination responses of many small seeded vegetables now grown in Europe may have changed little, if at all, during the period they have been under cultivation. This may imply that the possession by a species in the wild state of a germination response which is already incidentally adapted to the requirements of cultivation is a major factor influencing the probability of its adoption as a cultivated plant by mankind. An issue which it is essential to determine in any interpretation of the significance of germination strategies in relation to plant distribution, is whether the responses identified are adaptive, resulting from selective forces exerted specifically by the local environment, or adapted, representing the tolerance of an existing set of responses to a variety of conditions.

Comparisons between the germination responses of different populations of a species frequently appear to be put forward as evidence of the existence of adaptive changes. Reports on *Typha* spp. by McNaughton (1966), *Tsuga canadensis* (Stearns & Olson, 1958), *Amaranthus retroflexus* (McWilliams et al., 1968), *Avena* spp. (Thurston, 1957, 1962), *Impatiens* (Jouret, 1976), *Arabidopsis* (Napp-Zinn, 1976), amongst many others, have established the existence of differences in germination responses which could be related to variations in natural origin: in most cases the observations simply recorded the existence of variation between populations. Some recent work such as that by Makhnev & Makhneva (1979) on *Betula* spp. in which optimum temperatures for germination were found to be lower in provenances from higher altitudes in the Urals, and by Thompson (1975) on *Silene dioica* in which different populations displayed varying levels of dormancy at 11 °C, depending on their origin, provide results in which there is sufficient detail to relate to a broad view of the overall germination responses of the species involved.

Similarly, polymorphisms both in morphological characters of the seed coat and physiological aspects of the germination response have been related to environmental conditions in a number of species including *Chenopodium alba* (Williams & Harper, 1965), *Spergula arvensis* (New, 1958, 1959) and *Atriplex heterosperma* (Frankton & Bassett, 1968). Such observations have included the identification of klines in which particular morphological characters display progressive changes in frequency apparently related to climatic variations within the range of the species. It has been shown that the various forms of such species possess different germination characteristics. Edgar & Springfield (1977) have analysed responses of *Atriplex obovata* in which they separated seed samples into five groups depending on size and colour and were able to demonstrate differences in responses to temperature from one group to another, based almost entirely on proportions of seeds which remained ungerminated either in light or continuous darkness. Ungar (1979) has shown that large seeds of *Salicornia europaea* possess greater salt tolerance than small ones, and germinate more freely under any condition; in this species dimorphisms depend on the position on the plant of the flower which formed the seed. In most other species their origin appears to be more random, and wide variations in proportions of different types of seeds may occur between populations. Even in such examples, however, the strategy still persists of producing both, or a number of different, types of seed or physiological responses throughout the range of the species and variation is most commonly found in the proportions of each type present rather than in the replacement of one type by another. In a sense therefore the strategy of the species is to maintain variability throughout its range and this may appear as one of the most significant features of the germination response of the species involved. An interesting example has recently been identified in a particularly subtle form in *Stellaria meadia* by Vegte (1978), who has suggested that within Europe, at least two discreet populations may co-exist, frequently occurring in mixed communities but distinguishable by small differences in their phenology, and considerable differences in their germination behaviour.

Many experiments which examine variations between populations are done or presented in forms which make it hard or impossible to assess the significance of differences found in relation to the response of the species over a wide range of conditions; or to identify components of the variations found in a way, which would be essential, for example, in order to interpret the situation in *Stellaria media* noted above. Such reports, more often than not, identify quantitative differences between populations but do not define the overall germination strategy of the species. There is the possibility, since most workers expect to find differences, that when found they are reported with a prominance which may not be accorded to other experiments or conditions which fail to reveal variations between populations. Helliwell & Harrison (1978), in studies on the germination, growth and development of *Acer pseudoplatanus* and *Betula verrucosa* identified very few differences in the germination properties of different populations of either species, but did find major differences in the growth responses of seedlings to different types of soil. In a study on aspects of the control of germination, Semenza et al. (1978) working with eleven populations of *Verbascum thapsus*, found that tests at ten different temperatures between 2 and 45 °C displayed statistically significant variations in the responses of populations tested at particular temperatures, but markedly similar qualitative patterns in relative responses to different temperatures.

Other studies on groups of populations or cultivars have produced similar patterns in which quantitative differences appear to be laid over qualitative similarities. Cultivars of *Lactuca sativa* (Thompson et al., 1979) differed in details of their level of tolerance to high temperatures, but all displayed extremely uniform responses to temperatures below 20 °C, in their susceptibility to induced secondary dormancy, and in their capacity for rapid germination at low temperatures. Populations of *Agrostemma githago* tested at different temperatures displayed closely similar results to those of the cultivars of lettuce, and demonstrated the way in which the natural germination responses of this species have been retained through several thousand generations in cultivation (Thompson, 1973). In the case of this cornfield weed cultivation was inadvertant, so that no deliberate selective pressures were involved. Even so it would appear that the arguments suggesting stability of germination responses in *A. githago* can be applied with equal force to cultivars of *L. sativa*.

Wild species may possess complex reproductive strategies in which a balance of options is maintained without rapid selection in favour of one strategy or another. McNamara & Quinn (1977) have recently described ways in which resources are allocated in relation to the reproductive strategy of populations of the grass *Amphicarpum purshi* which is a pioneer species in secondary successions. This species produces dimorphic (aerial and subterranean) caryopses of greatly different size and competitive capacity, and has two modes of reproduction involving either cross or self-fertilisation. This variability is maintained throughout the life of a population and during invasion of new habitats, though the proportions of different types of caryopsis contributing to successful seed establishment varies greatly with conditions. As the authors comment 'the dimorphic mode of reproduction ... includes one form maximally suited for high probability of reproductive success in an unchanging environment ... and another form which has a lower probability of survival but maintains greater variability by gene flow between populations, or, more likely individuals'. The species exhibits in a vivid way a strategy based on continuously maintained variation which persists in the face of what appear to be strong pressures tending to selection for one form rather than another.

Situations also occur in which corresponding germination strategies appear to have been maintained by groups of closely related species throughout the course of speciation and occupation of a variety of more or less diverse habitats. Thus within certain sections of large genera such as *Primula* (Thompson, 1970a) or *Gentiana* (Thompson, 1969), similarities in germination behaviour may be correlated with broadly similar or analogous habitats, geographical ranges or phenologies of the species being compared, so that selective pressures tending to change might be expected to have been relatively small. Oesau (1975) in a study of seed germination and root development of six different species of *Melampyrum* demonstrated strikingly similar responses in all six, five of which displayed optimum germination between 2 and 10 °C corresponding to field conditions during October and November, and the sixth a similar minimum, but germinating up to 14 °C. On the other hand studies on seven species of *Alstroemeria* (Thompson et al., 1979) distributed in the southern Andes in a variety of habitats ranging from the coastal plain to montane areas have shown very strong affinities between the germination strategies of all seven. Each responded to a conditioning treatment at high temperatures, and each germinated optimally at low temperatures (6-10 °C) with some degree of dependance upon degradation of the seed coat. It would appear that these species have retained ancestral germination strategies which have persisted throughout the processes of speciation, and even establishment within distinct geographical and climatically varied ranges.

It is evident therefore, despite reports of variations in germination responses between populations of differing origin, and of changes in dormancy restrictions in certain cultivars, that there are examples in which a major feature of the germination strategy is its stability over very many generations. This raises questions about how such stability may be achieved, and how it is possible for a character such as germination which appears intrinsically so subject to selective pressures, and which is frequently based on polymorphisms apparently ready-adapted for selection, to remain substantially unchanged even in some cases through the course of evolution of a group of species which themselves come to occupy distinctively diverse habitats.

Maintenance of populations

The previous section has, deliberately, emphasised the stability of germination responses. The recent literature provides examples of variations in the responses of closely related species which appear to correlate with climatic or other features of their natural distribution - suggestive of adaptive responses. For example Sonia & Haslehurst (1978) identified well defined variations in the temperature responses of three species of *Banksia*, which occur under different conditions along the coasts of eastern Australia and which could be correlated with particular climatic conditions characteristic of the range or habitat of each species within the area. Similarly, Ware & Quaterman (1969) and Bouchard & Franz (1977) defined differences in requirements for moist cold treatment, light and alternating temperatures between species of *Talinum* occurring in similar habitats in the south-eastern United States, which draw attention to differences between the species.

The very great reduction in numbers which occurs each generation between dispersal and the maturation of the surviving individuals which provide the following seed crop, might be expected to render the processes of germination particularly susceptible to adaptive change. Such a pattern would provide every opportunity for stringent selection based on the genotypes of successful seedlings, leading to rapid changes in the character of local populations in response to changing conditions, or new habitats encountered during natural colonisation or the artificial distribution of the species by man, particularly as a weed or cultivated plant.

The anatomical composition of the seed which encloses an embryo with a particular genotype surrounded by tissues, which possess the maternal genotype, would diminish effects of such selection. However, given the close relationship which exists between embryo and parent this complexity would do no more than reduce the rate of selection, rather than inhibit or disrupt it. There is also evidence that under certain conditions, such as those imposed by Morley (1958) on *Trifolium subterraneum*, changes in some germination responses at least can be quite rapid.

Harper (1977) has commented that population biology is concerned with the numbers of organisms and the consequences of those numbers, and Salisbury as long ago as 1942 remarked that the one to one, or one to two (in dioecious species) relationship between seed parents and mature offspring in a stable population has been quoted 'ad nauseam', but it still remains the case that, under natural conditions, thousands, tens or hundreds of thousands of seeds are regularly produced in order to secure the maturation of very, very few mature individuals. Sharma & Gopal (1978) in a very recent publication have commented that although individuals of species of *Typha* produce billions of seeds, seedlings have rarely been reported in nature, yet this is one of the examples, following work by McNaughton (1966), which provides data on the nature of variation in germination responses between populations. Any assessment of the significance of particular germination strategies displayed by a species, must clearly be made with an awareness of the odds against survival that apply to any individual seed.

Salisbury (1942) has also pointed out that populations, which depend on seed as their main reproductive mode are seldom stable from year to year, but tend to experience large fluctuations in numbers, and he has suggested that such fluctuations may be positively beneficial as a means of disrupting patterns of predator and pathogenic infestation. In practice, the seed may be seen as one way of maintaining the presence of a population, whatever its level may be within a particular area, but not particularly as a means of guarding against extinction of the population. It is also clear that the seed not only fulfills a function within the phenological development of the plant, which must harmonise with all other stages of development, but that it also fulfills a defensive role by virtue of its abundance and expendability in relation to predators, through which a very few individuals survive as a result of the sacrifice of the great majority. Recent work has developed to a considerable extent our understanding of the role of predators and their patterns of predation, and of the significance of germination strategies in relation to the phenology of different species, and have made it possible to begin to relate the significance of seed dispersal and germination strategies to a broader ecological context.
Salisbury proposed that the reproductive capacity of a species is correlated with the risks of mortality - and this concept must be accepted since the two form the opposing forces that result in the balance of a population, at least in a post hoc concept though there may be no reason to believe the two are linked by design. It follows that a species will persist in any location within which seed production happens to exceed mortality, and is incapable of persistence when this does not apply. A recent study by Yeaton (1978) on ecological aspects of the germination strategies of two groups within the genus *Pinus* illustrates the subtlety of differences in reproductive strategy that may define survival. The groups involved are members of the sub-genus *Haploxylon* on the one hand which possess no apparent tolerance to water shortage, and on the other the subgenus *Diploxylon*, which display extensive adaptations to water stress. Consequently members of the two sub-genera occupy different microhabitats within common geographical areas. It was found that species of *Diploxylon* displayed reproductive characters such as early maturity, a short interval between large seed crops, large numbers of cones per tree, and small seeds which favour chances of emigrating, and finding suitable niches in which to germinate and develop. Species of *Haploxylon*, however, possess a contrary reproductive strategy which provides for greater competitive ability and maximum vegetative growth under more stable and consistently moist conditions. Within this context germination is very clearly only one process amongst many which define chances of survival of a population.
It is also evident that changes in germination - of any kind - are likely to be advantageous only in special circumstances, or when accompanied coincidentally by changes in a large number of other attributes which contribute to the phenology of the species. Jouret (1977) in a study on a number of species of *Impatiens* naturally distributed over a wide geographical range has related germination processes, including levels of dormancy, to the chorology of the

species, and published data which clearly illustrate the need for harmony between the seasonal succession of local climatic factors and the relative needs of different phenological stages, including the timing of germination. The requirement for such harmony may be seen particularly clearly in populations from areas with well-defined seasonal changes which include periods of very severe stress. For example, annual species from Mediterranean areas of Europe have no practical chance of survival if they germinate in response to transient periods of summer rainfall, and Thompson (1970b) has compared the germination responses of species of Caryophyllaceae from this area with those from other parts of Europe - particularly western oceanic regions where the balance of advantage between germination at one season rather than another may be a very fine one. Fundamental changes not only in germination responses but in other aspects of the course of plant development may be required to enable extensions of range in such circumstances, and Morley & Katznelson (1965) have discussed the factors which contribute not only to the survival and spread in Australia of *Trifolium subterraneum*, but which also restrict its distribution within particular well-defined limits. They have considered this as a naturally well-adapted colonizing species, and have then shown how climatic, edaphic and biotic limits interact to define the area of colonisation by the species in ways which appear to leave little scope for a significant contribution leading to any extension of range as a result of changes in the germination responses of the seeds.
An interesting case of survival based on a non-specialist germination strategy has recently been reported by Augspurger (1979) with reference to the Panamanian shrub *Hybanthus prunifolius*. This species grows naturally in an area subject to drought interspersed with periods of unreliable and usually inadequate rainfall. The species responds to the erratic occurrence of adequate moisture by shedding seeds which possess no dormancy mechanisms and no protection against germination in response to inadequate rainfall. Augspurger argues that this is a successful strategy since seed dispersal occasionally coincides with conditions which will maintain growth and survival, and he suggests that any more specialised response would fail to take advantage of erratic favourable conditions, or would require strong anti-fungal or anti-predator defences, which would be an uneconomic allocation of plant resources.
Kirkpatrick & Bazzaz (1979) have drawn attention to the role of fungal pathogens in the establishment and survival of seeds and seedlings, in experiments on four colonizing annual species; *Abutilon theophrastii*, *Datura stramonium*, *Ipomoea hederacea* and *Polygonum pennsylvanicum*. They found that pathogens were mostly host-specific, and mainly affected seed germination and the development of young seedlings, and that wide variations occurred in the level of resistence to infection between the species. Amongst this group of species which all grow within the same habitat, differences in strategy appeared in which some species reduced the minimum reproductive requirement for survival by the production of defences against fungal infection, others survived through the production of sufficient seeds to ensure the survival of some in spite of the risk of fungal infection.

In any situation the penalty for germinating at a selectively unfavourable season of the year, or under other adverse physical or biological conditions, is an increased chance of mortality before maturity. But, provided this increase does not exceed the natural reproductive capacity of the population it does not follow that the population will be unable to establish or maintain itself, still less that any change in germination response is essential for survival. It should also be borne in mind that the level of mortality derives from a great variety of events some of which might be reduced in severity by selection for particular genotypes, such as frost tolerance, pathogen resistence etc., but many of which are almost entirely random in nature depending on the precise location in which a seed finds itself and the relevant patterns of predation by vertebrates and insects. Zlobin (1977) has related ways in which the germination of seeds of *Pinus sylvestris* and *Picea abies* is affected by the internal physiological condition of the seed and has also linked this with a large number of physical and mechanical properties within the soil micro-habitat, all of which must be maintained coincidentally within certain limits before successful germination can occur.

Harper (1977) has suggested that the numbers of seedlings of a species which establish successfully is a measure of the number of 'safe sites' afforded by their environment. In this context a 'safe site' is one which provides the stimuli to break dormancy, the conditions required for germination to proceed the resources necessary to maintain growth and the absence of overwhelming destructive elements in the environment such as predators or toxic chemicals. Catastrophic events such as fire may create numerous safe sites within vegetation, though often only very temporarily, and their effects have been widely studied including papers by Archibold (1979) who examined the part played by seeds in the regeneration of forests in northern Saskatchewan, and by Cabiaux & Devillez (1977) who related successful establishment of *Betula pubescens* to the physical nature of the soil surface and particularly its texture and hydration. They linked these qualities with the high level of natural regeneration by birch which characteristically follows forest fires.

Grazing animals also have a major influence on the number and availability of safe-sites by creating them, often as a result of trampling or degrading existing vegetation, by modifying the nature of existing flora or by altering dispersal patterns of the propagules of the natural vegetation. In mediterranean areas of the world seasonal regeneration may depend heavily on annuals and Gramshaw & Stern (1977) have studied the regeneration and persistence of the annual grass *Lolium rigidum* in Western Australia, under conditions of intensive stocking with sheep. They found that high proportions of seeds were dislodged from the seed heads during heavy grazing and that it was these shed seeds which largely contributed to the new emergent population of seedlings the following autumn. Seeds retained in the old inflorescences were not in a position to make use of early rainfall at the beginning of the growing season, and either failed to germinate at all, or did so later under unfavourable competitive conditions.

Germination strategy and environment

The position of the seed in relation to the soil surface may critically define chances of germination and in a notable contribution on this topic Harper et al. (1965) showed that irregularities on the soil surface not only substantially affected the prospects of seedling production in three species of *Plantago*, but affected each species in a characteristically different way - arising possibly from differences in mucilage production, and the effectiveness of the soil/seed interface in each species. Blom (1978) has reported on a wider range of *Plantago* species including results obtained in field experiments in which different species displayed well-marked preferences for particular habitats based on successful seedling establishment. *Plantago lanceolata* and *P. canopus* were favoured by open and dry habitats; *P. lanceolata* by tall, dense layers of vegetation. *P. major* spp. *major* was most successful on moist and open sites with compacted soil, and *P. major* and *P. coronopus* displayed opposite mortality curves on the surfaces of trodden paths. The results extend the significance of the earlier results obtained by Harper et al. (1965) and also emphasise the almost random nature of the way successful germination and establishment depend on the precise location within particular definable situations, in which the seed finds itself. Waite & Hutchings (1978) found in laboratory tests, and in experiments using semi-natural substrates that seed of *Plantago coronopus* germinated and established most successfully when sown in clumps and produced lower proportions of seedlings when scattered at random. Subsequent experiments in the field (Waite & Hutchings, 1979) failed to confirm this effect, but did show that the recruitment of seedlings into the population was limited by the number of adult and seedling plants already present, suggesting that numbers were in fact limited under these conditions by the number of 'safe sites' and their saturation by established members of the population. Noble & Whalley (1978) have shown the significance of 'safe sites' as a major factor regulating seedling survival of *Nitraria billardieri* in Atriplex scrubland in Australia, in a report which showed that ingestion of seeds by emus and their clumping in faeces greatly increased chances of reproductive success. Oomes & Elberse (1976) have compared germination responses in six species of herbs occurring in grasslands in the Netherlands in relation to micro-sites, which differed in the availability and consistency of water content. They identified differences in germination behaviour which depended largely on variations in rate and facility of water uptake, according to the form of the seed or the presence of mucilage, and they related these results to prospects of establishment of seedlings under the conditions of varying water stress typical of natural situations.

It follows that the essential requirement for the survival and maintenance of a population of a species in a particular location depends, not upon the production of a relatively few precisely adapted individuals, but on a cadre of some size, which, taking climatic and other hazards of one year with another results in an adequate number of established seedlings, year by year, decade by decade and century by century. This is not unlike a concept expressed by

Stebbins (1958) in relation to highly heterozygous out-crossing species, in which he points out that when a population produces progeny in far greater numbers than are able to survive, then the advantage will go to those lines which produce at least a small number of highly fit individuals even though a high proportion of their siblings may be poorly adapted to the same situation. It is logical to argue that heteromorphic seed responses would produce an analogous situation in which the 'right' responses (whether resulting from genotype or random opportunity) of a very few individuals contribute to the survival of the population and the fate of the great majority of the population, is of no significance.
The germination strategy and the detailed responses of any particular individual, or group of individuals, seldom absolutely determine whether or not the population can maintain itself, but, by interacting with the environment, it raises or lowers the odds against survival, taking them above or below the reproductive capacity of the species involved in that particular location.

Germination strategies and the natural environment

There are many situations in which the prospects of survival from seedling to maturity are better at certain seasons or in particular locations than others, and in a number of reports the results of tests on germination responses have been correlated with the natural environment of the species involved. Thus Young & Evans (1979) compared two related species of *Salsola*, and found generally similar germination characteristics. However, *S. paulsenii*, which occurs in the most arid positions of severely degraded communities differed in a number of respects from the more mesic *S. iberica*, all of which contributed to survival in arid environments. Experiments on *Hyacinthoides non-scripta* (Thompson & Cox, 1978) produced results which showed how the germination responses of the species were well adapted to its normal occurrence as a perennial growing naturally in the deciduous woodland of oceanic areas of western Europe, but which incidentally, also provided an interpretation for the relatively narrow geographical range of this species resulting from the inability of its seedlings to withstand severe winter cold, or summer drought. Mott (1978) has compared the dormancy and germination responses of five grasses native to savannah woodlands of the Northern Territory of Australia, and related patterns of dormancy, germination and auto-immolation within the soil, to seasonal patterns of temperature and rainfall and the frequent occurrence of grassland fires in the area. Similarly, Baskin and Baskin have reported on the germination of a number of different species and linked their germination responses to the ecology and phenology of each species as a natural component of the north American flora. These species include *Sedum pulchellum, Asclepias syriaca, Rumex crispus, Cyperus inflexus* and *Aster pilosus* (Baskin & Baskin, 1977a, 1977b, 1978a, 1978b, 1979). They have been particularly successful in defining the responses of seed at the time it is shed in late summer and autumn, and then following the course of the conditioning treatments which alter responses during winter chilling or summer warmth. In this way they have provided interpretations of the patterns of germination in relation to the natural seasonal succession of the areas in which the populations grow. There are now numerous reports of ways in which seeds use local environmental stimuli to bring about germination at a particular season, and some correlate such patterns with overall aspects of the phenology of the species involved. These include reports on two forbs occurring in *Artemisia* sagebrush rangelands (Young & Evans, 1979), effects of sowing dates on the survival of two species of *Trifolium* (Evans et al., 1976), persistence of the woody weed *Baccharis halimifolia* (Panetta, 1979), responses of the salt tolerant halophyte *Salicornia europaea* (Ungar, 1979), reproduction of *Gentiana crinita* in eastern USA (Farmer, 1978), and factors affecting the very infrequent germination of the hard-seeded *Rubus chamaemorus* in Newfoundland and Labrador (Warr et al., 1979). These reports which cover a broad diversity of flora and geographical situations provide considerable information on the nature and significance of the responses which control germination under natural conditions. They do not provide definitive evidence for selective changes, nor indications of rates of change, which might produce adaptive populations of a species better fitted to the local environments in which they occur, and so almost always fail to answer the question whether the response observed is adaptive or adapted.

Factors influencing the course of adaptive changes

Superficially it might appear that advantages would result from selective changes which caused higher proportions of seedlings to be produced at seasons favourable for survival and final maturation. This may be so, but it is not inevitably so. In the first place many environments are unpredictable from year to year, so that there is no unequivocal best season for germination - or best set of conditions to which to respond. Under these conditions diversity of response, and mechanisms for the long term maintenance of germination and other polymorphisms provide essential safequards against the short term selective changes which would otherwise result from an atypical succession of seasons with a tendency towards consistency - whether this occurred as a series of unusually cold winters or arid summers.
Secondly the numbers of mature individuals present within a population may not be limited by numbers of seed or the numbers of seedlings produced, but by the presence of unoccupied 'safe sites' which define the number of plants which establish successfully. Even when this is not the case increases in the numbers of individuals of a species present in a particular area may itself lead to increased predation or severity of pathogenic infestation, and so increase odds against survival. The situation is illustrated in tree species by the occurrence of 'mast' years in which large crops of seeds and large numbers of emergent seedlings, frequently lead to extreme levels of predation, so that survival even of a minute proportion of individuals may occur only in exceptional years when for one reason or another seed production outstrips the rate of predation. This situation includes many analogies with patterns of seed and seedling production in herbs or shrubs in which seedlings

may occur in 'flushes' or in relative isolation at different locations or seasons. The ways in which predators and seeds or seedlings interact have been the subject of a number of reports, which focus interest particularly on the foraging behaviour of ants (Davidson, 1977a, b) and Davidson & Barbour (1977) and mice (Martell, 1979), and patterns of infestation by bruchid beetles. Ashton (1979) has described how ants in mature *Eucalyptus regnans* forest may destroy two thirds or more of the seeds produced in normal years and are a major force in the prevention of seedling regeneration. It is suggested that the reduction in the numbers of ants and the disturbance to their foraging patterns which result from fire may critically contribute to successful regeneration by this species. Foraging patterns of ants in New Mexico reported by Whitford (1978), displayed clear-cut seed density-dependant effects in which those species which shed large numbers of seeds at a time were preferentially selected. Ants have also been found to have preferences for seeds of particular species, which may markedly affect the chances of establishment and success of heavily predated taxa. Reports by Johns & Greenup (1976a, b) have provided measurements of rates of predation of seeds of *Trifolium repens, Medicago sativa, Festuca arundinacea* and *Phalaris tuberosa* artificially distributed in pastures in northern New South Wales, and emphasise the preferential predation of *P. tuberosa*. Similar indications of preferential predation, following the provision of exotic seeds of cultivated plants including oats, and sunflower, have been reported by Sullivan (1978, 1979) for deer-mice feeding on seeds of *Tsuga canadensis*. Differential preferences arising from changes in foraging patterns by ants and mice, and very often based on the relative abundance of a number of different species, are much less common in relationships between seeds and other insect predators, which are frequently highly host-specific. Janzen (1977c) in a study of predation by a bruchid beetle on the large leguminous tree *Pithecolobium saman* found no evidence that increased availability of fruit increased predation - but rather the reverse due to saturation effects on the population of beetles present. Hare & Futuyma (1978) examined patterns of predation by two different monophagous insect species on ten populations of *Xanthium strumarium*. They found that populations differed from one another particularly in burr morphology and that the form of the burr could confer resistence to both insects simultaneously; other characters including chemical composition provided resistence to one or the other. The results imply that plant resistence to insect predators may be interdependant and that defences against one may or may not confer resistence against another - and could in some circumstances even lead to enhanced susceptibility. Thus strategies leading to improved possibilities of survival of populations of a species may be based on greater protection against predators, more accurate timing of germination or greater seed productivity. But all three strategies interact and confer advantages and disadvantages, which directly relate their success to the demands they make on total available plant resources. The benefits of evolutionary change away from established patterns of germination towards others apparently better fitted to a local environment, measured by the capacity to provide more seedlings at a favourable season, may also be constrained by a third factor. It is recognised that seeds may contribute effectively to the maintenance of populations by remaining in the soil seed bank, and there are now numerous well-authenticated reports such as those by Ødum (1965) which document the survival of seeds of many species for very long periods in this way as a normal condition. Thompson (1979) has recently reviewed the significance of this strategy in relation to various environments concluding that levels of buried seed within the soil are correlated positively with disturbance and negatively with stress. Consequently the seed bank becomes a productive strategy in conditions such as cultivation, or natural persistent soil disturbance where selection favouring buried seed is intense, but is of minimal significance in areas such as tropical rain forest or arctic/alpine habitats. Patterns of dormancy in seeds, which lead to the accumulation of individuals within the seed bank, therefore have different significance in different locations; according to whether the main function played by the bank is a short-term, or seasonal, conservation of a flora regenerating annually, or a long term insurance against spasmodic and catastrophic changes in the composition of the flora (Thompson & Grime, 1979). The seed bank plays its part most effectively in situations where colonisation and changes in range are most active, and reports such as those of Charlton (1977) on pasture legumes, Bartolome (1979) on components of annual grassland in California, and Jones & Evans (1977) on numbers of seeds of species of Leguminoseae in sub-tropical pastures are typical of many in which it can be seen that the presence of the bank plays a major part in maintaining the diversity of plant representation within a community. It is particularly noteworthy that the composition of such seed banks practically never reflects the constitution of the existing established flora in relative abundance, and may contain species which have disappeared during early successional stages in the development of the vegetation. The existence of the seed bank and its role as a means of maintaining diversity, especially under disturbed conditions, suggests that, in some circumstances, quantitative changes in the proportions of dormant seeds could be a much more effective form of adaptation to local conditions, than even quite major changes in germination strategies based on responses to temperature or light. Laboratory experiments on the germination of populations of *Silene dioica* from Europe and of other perennial species in the family Caryophyllaceae (Thompson, 1970b, 1975) have drawn attention to variations in proportions of dormant individuals as one of the major components of variation between populations, suggesting that partition between the emergent population and the seed bank is of adaptive significance in relation to geographical distribution.

Reports have also stressed the role of the seed bank as a mechanism which increases the genetic base of a population by mixing cohorts of one year with another, thus mitigating selective pressures arising from short-term sequences of unusual conditions. The opportunities for miscegenation of this kind are exemplified in a recent report by Del Tredici (1977) on the persistence of living seeds of *Comptonia peregrina* for a minimum of seventy years in a forest community, and of other reports by Floyd (1976) and Purdie (1977) on the role of the seed bank during regeneration by many species, including both grasses and forbs, after destruction of long-established forest. Although the long-term survival of seeds becomes most apparent after the destruction of established forest, there are many reports such as those referred to earlier on grasses and legumes, and by Valk & Davis (1976) on the composition and nature of seed banks in glacial marshes in Canadian prairies, and by Hodgkinson et al. (1980) on the seed bank of scrub woodland in New South Wales, which indicate its role as a source of continuous recruitment to the plant population of many communities. Baskin & Baskin (1978c) have described how these effects can be particularly critical in the case of the narrowly based endemic species, *Leavenworthia stylosa*, in which the presence of individuals from at least seven different previous harvests may be a major factor favouring genetic stability within a population. They contrast this situation with that of certain *Clarkia* spp., which Lewis (1962) has shown fail to produce a soil seed bank, and which appear to be unusually prone to genetic disruption leading to an active pattern of speciation. However, other species such as *Cerastium atrovirens* and *Vulpia membrenaceum* which Harper (1977) believes also fail to produce a seed bank reserve, do not apparently exhibit unusual genetic instability, and are not prone to active speciation.

Variation in germination behaviour between populations

Examples have been cited of situations in which the germination strategies of different populations of a species display little variation from one geographical location to another. The arguments presented in earlier sections suggest that, given a species with a fixed strategy, the odds against survival of an individual seed would vary from one geographical or edaphic location to another. This means that the minimum reproductive level necessary to maintain the population would be less in favourable parts of the range or under relatively unstressed conditions, implying that survival in, or extensions in range to less favourable areas, would depend either on changes in germination strategy, accompanied by other adaptive changes in phenology and, possibly, on protective defences against predators or pathogens, or an increase in seed numbers or seed size to match or reduce increased odds against survival. However, there are probably as yet, no species, about which enough is known of the whole range of factors affecting reproduction and survival of populations, to enable accurate comprehensive definitions of favourable or unfavourable conditions to be made.

Harper et al. (1970) in a review of the significance of seed size have discussed the conditions under which this character is variable, but recognising as Salisbury (1942) has pointed out, that increases in seed size normally correlate negatively with seed numbers, since these two components of seed production interact competitively for limited resources. Nevertheless, adjustments in seed size correlated with climatic conditions or competition have been reported, for example by Kandya et al. (1978) who found that seeds of *Pinus caribaea* produced by populations in drier areas are larger and heavier than those produced in more mesic situations, and by Jurka

& Benneweg (1979) who reported considerable variation in seed size both within and between populations of *Capsella bursapastoris* from different parts of Europe. Plants in severe climatic conditions produced larger seeds than those in favourable ones, and there appeared to be a high level of genotype × environment interaction in seed size, which provided the plants with a high degree of phenotypic plasticity. In many situations species would be incapable of increased reproductive capacity under conditions of greater stress and would have to opt either for increases in seed size or seed number. However, extensions of range do not inevitably involve increases in those stresses which directly affect reproductive capacity - for example mediterranean annuals which also occur as winter annuals in western oceanic Europe as weeds, ruderals, or other components of the flora of disturbed ground may, provided they have sufficient cold tolerance, grow under conditions which are more productive than those in their original range - and in such cases greatly increased seed productivity may be a part of the strategy developed during range extension.

There is accumulating evidence that the phenotypic plasticity possessed by many species (Bradshaw, 1965) involving levels of post-harvest dormancy, absolute limits of response to temperature, responses to conditioning treatments and other aspects of the germination response, may modify overall germination strategies in ways which reduce unfavourable odds against survival on the edges of established ranges.

Within a particular locality a range of different germination strategies from one species to another such as those described by Hnatiuk (1979) in a study of ninety species collected on Aldabra Atoll in the Indian Ocean, would not only partition the flora into a number of separate essentially non-competing components, but would also produce different odds against survival, and so give rise to variations in the minimum reproductive capacity required for the continued maintenance of populations. Grubb (1977) has reviewed the significance of the regeneration niche in the maintenance of a species rich flora, and Went (1955) reported the different strategies of desert annuals which provide excellent examples of the ways in which species may partition a particular environment. He distinguished four sets of species within the Mojave and Colorado deserts which could be separated by their germination responses - one set germinating at high temperatures after summer rains which occur irregularly at intervals or five to twenty years. A second group germinates at relatively low temperatures at the height of the relatively regular winter rains, a third at intermediate temperatures in early winter, and the fourth, opportunist in nature, with no set pattern in the timing of their germination. Each strategy placed different demands upon the species involved, producing different odds against the survival of individual seeds within which particular species would fit depending on their reproductive capacity, and the tolerance of their phenotype in relation to the environment. Recently Raguse (1977) has reported observations on germination of winter annuals in response to heavy summer rain in Californian foothill rangelands, which he describes as a 'once in a hundred years' event which draw attention to the stability of such seasonal responses. In spite of highly favourable conditions for several days, high temperatures inhibited the germination of all but a few of the annual species, and germination when it did occur was greatly favoured by the presence of micro-sites - providing prolonged cool-moist conditions such as the footprints of cattle on north facing slopes.

Species which occur in close proximity, and in broadly similar habitats, may possess totally different strategies, leading to dependance on a variety of regeneration niches as suggested by Grubb (1977), and greatly different minimum reproductive capacities. Slade & Causton (1979) have reported results of tests on the germination of thirteen species of herbs which grow amongst deciduous woodlands in the UK in response to five treatments potentially able to break dormancy. The ability of these species to germinate without treatment, to respond to one or more of these conditioning treatments, or to fail to germinate under any of the experimental conditions set up, varied very widely. Many showed great sensitivity to the presence or absence of particular conditioning treatments, sometimes with little apparent logic from one combination to another. Apart from variations in their innate strategies, the positions in which seeds repose after dispersal introduce other variables controlling the probability of germination arising from temperature conditions, availability of water and exposure to light. One of the major constraints which operates frequently even on seeds with few other germination restrictions is the presence of an overhead leaf canopy. Black (1969) and Taylorson & Borthwick (1969) have drawn attention to analogies between laboratory experiments using red and far-red lighting regimes to investigate the phytochrome photoreceptor system in seeds, and the effects of the light screening properties of the leaf canopy. A recent broad screening of more than two hundred herbaceous species by Gorski et al. (1977, 1978), has demonstrated how widespread this phenomenon is. Almost all the wild species occurring naturally in open habitats were highly sensitive to the inhibitory effects of leaf canopies, but the common weeds of cultivation produced more diverse responses. Twenty seven species from calcareous grassland were the subject of investigations by Silvertown (1980) who found that a very high proportion displayed induced dormancy under leaf canopy conditions, to the extent that germination was significantly below that of dark-grown seeds in seventeen of them. Vazquez-Yanes (1980) has reported similarly inhibitory effects on the germination of two species from Mexican rain forests. Apart from direct inhibitory effects on germination by leaf canopies there are also reports that such conditions may induce light requirements in species such as *Bidens pilosa* (Fenner, 1980a, b), or affect the progression of responses to conditioning treatments such as winter chilling in *Aster pilosus* (Baskin & Baskin, 1979).

Variations in reproductive strategy, seed production rates and overall phenology between two species from similar habitats are well illustrated by *Digitalis purpurea* and *Hyacinthoides non-scripta* which both occur as components of the deciduous woodland ground flora in western oceanic Europe, but which possess distinctively different phenologies and germination strategies. Salisbury (1942) has reported average seed production rates of about 85 000 seeds per plant from *D. purpurea* compared with 95 seeds each year from

H. non-scripta; even allowing for the greater longevity of the latter there is no doubt of its lesser productivity. Levins (1968) has pointed out that the cost of ability to function at a wide range of environments is obtained only by reducing efficiency at the optimum, and differences in the germination strategy of these two species illustrate this very clearly. *H. non-scripta* has a highly specialised response (Thompson & Cox, 1978) based on particular temperature conditions which result in precise control of the season of germination. This is a highly successful strategy in western oceanic areas with generally mild winters, but has the disadvantage that it results in low chances of survival of the winter germinating seedlings in colder continental winters, or in areas with dry warm summers in parts of southern Europe. *D. purpurea* displays no post-harvest dormancy, it is able to germinate over a wide range of temperatures, and the primary factor restricting germination is the presence of a leaf canopy; a secondary response being the induction of dormancy by low temperatures. The species occurs naturally in western Europe and, unlike *H. non-scripta*, displays marked colonizing ability outside its natural range; though very frequently this success is ephemeral. Its germination responses display the wide environmental tolerance characteristic of the general purpose genotype which Baker (1965) has described for weedy species, a concept which Cole (1977) has also taken up in a suggestion that similar characteristics contribute to the successful colonizing of West African swamps by the introduced grass *Panicum laxum*. However, the tolerance of the germination response of *D. purpurea* is not matched by similar tolerances in other aspects of the plant's genotype, and under unfavourable conditions, particularly of competition, post germination mortality produces odds against survival which exceed even the high reproductive capacity of the species.

Examples of other highly specialised responses tailored to particular situations are found in many environments - a striking one is the adaptation to fire shown by some species of *Pinus, Eucalyptus* and *Banksia* in which seeds remain within the fruit bodies until released by fire to colonise burnt-over areas. Gill (1976) has described the ways in which temperature and moisture interact to rupture the follicles of *Banksia ornata*, which normally open after fire, but are capable of dispersing seed under some other conditions - for example, following the death of the parent plant. Siddiqi et al. (1976) have described conditions favouring the release of seeds, germination and establishment of two species of *Banksia* in response to fire, in which they report rapid germination after release of seeds with no specialised germination responses. The essential strategy in this case being the protection of the seeds within the hard follicles during the course of the fire.

Effects of seed dispersal on selective pressures

The point has been made that germination would appear to be a process particularly susceptible to selection as a result of the great numbers of seeds produced and the considerable advantages that may be conferred on the prospects of survival of seedlings as a result of germination at a favourable season of the year. It would appear that towards the geographical or environmental limits of the range of a species the presence of a few seeds with a genotype resulting in germination patterns slightly different to those of the majority, but better fitted for survival, could give rise to continuous selection for favourable characters and so, potentially to continuous extension of range. Similarly it appears inevitable that species with wide environmental ranges would differ in the ways in which seeds of different populations responded to environmental stimuli simply in order to maintain an appropriate strategy in the face of a variety of climatic conditions. Possibilities for such change would depend on the nature of the genetic control of germination, about which little is known and which will be considered later but would also be subject to other environmental and phenological constraints, in addition to those discussed previously.

Harper (1977) has commented that dispersal appears to consist of two distinct components, which in fact represent different aspects of the same situation. One being the processes by which populations invade a new area, island or continent: the other the ways in which an established population maintains itself within the constantly shifting 'islands' of suitable habitats within established vegetation. De Angelis et al. (1979) have produced a mathematical model representing the dynamics of species of trees dispersed by seed in a region containing patches of forest. Watkinson (1978) has analysed the role of seed dispersal in the demography of *Vulpia fasciculata* collected from coastal sand dunes in North Wales. He comments that a poorly dispersed seed runs the high risk of density stress from neighbours of its own kind, but is likely to occupy a site, which in other respects, is favourable for growth and successful reproduction. A well dispersed seed on the other hand runs a high risk of landing in an unsuitable site, but achieves relative freedom from intra-specific competition. Experimental studies suggested that this species was likely to spread as very locally dispersed seedlings in an advancing front from short infructescences, and as isolated colonists from tall infructescences. In the course of dispersal seeds of *Vulpia* are released from their parent plants approximately three months before they produce seedlings, indicating strong seasonal controls over the timing of germination. In such a situation well fitted germination response may play a crucial role in determining survival, and the pay-off for improvements in germination strategy would be a decreased reproductive requirement for survival. This would appear to be a potentially valuable pay-off particularly in situations involving extension of range or invasion of marginal habitats.

However, many of the factors which affect the chances of successful dispersal and seedling establishment place a high premium on the production of large numbers of seeds and a low one on improved adaptive responses. Factors of this kind have been summarised by Harper (1977) who suggests that, within regions widely occupied by a species, numbers of representative individuals may be limited for the following reasons: a) areas suitable for occupancy may be low, b) habitable areas may be far apart, c) carrying capacity of habitable sites may be low, d) the time during which sites are suitable for colonisation may be short, e) the time for which a site is

habitable may be short, f) colonisation and exploitation of carrying capacity of a site may be slow. Thus an essential requirement for successful establishment may be the production of large numbers of diaspores - simply to search out suitable sites - and almost irrespective of the nature of their germination responses.
The advantages of effective dispersal have led to numerous special mechanisms and symbiotic relationships with potential predators which have been and still are being extensively studied. Dispersal of seeds by ants is common amongst many plant families some of which commit valuable metabolic resources to the production of elaiosomes to encourage dispersal and reduce the probability of predation of the seed itself. Handel (1978) has reported on the evolution of ant dispersal mechanisms in *Carex pedunculata* and their role in seedling production and competitive capacity in the species. Dispersal by mammals including mice (Zhiryakov, 1977; Reichman & Oberstein, 1977), bats (Janzen, 1976) and squirrels (Stapanian & Smith, 1978) all involve elements of predation either on the seeds themselves in which high proportions may be lost to provide for the dispersal of a few, or on surrounding fleshy tissues. Similarly, dispersal by birds, particularly of fruit eating species may accomplish wide-range scattering of seeds - much of it in locations unsuitable for survival of the seedlings - but there are other examples, and one is reported by Howe (1977) in which species of birds remove fleshy fruit coverings and drop the seeds, uneaten, in situ beneath the parent plant.
These examples draw attention to some of the haphazard elements which govern regeneration by seed and which contribute to a situation in which the prospects of survival of any particular individual are exceedingly small. In such a situation even considerable genetic changes in germination responses would, in the most favourable circumstances, still be subject to very high odds against the survival of individuals bearing an improved genotype. In addition, the mixed parent/progeny genetic constitution of the seed, including the strong (often dominant) controlling influence of the maternally-derived tissues in many seeds, means that up to fifty per cent even of the resulting F2 generation could be without the altered genotype responsible for the improved response.

Alternative strategies

These arguments suggest that extensions in the range of a species may be determined to only a relatively small extent by changes, arising from specialised adaptations to the conditions under which seeds germinate. They may arise with greater effectiveness from simple changes in the proportions of dormant seeds remaining in the soil seed bank to ensure dispersal in time, or from increases in seed production rates to maintain a positive balance between seed production and mortality and improve the chances of dispersal in space.
The significance of the soil seed bank as a means of survival and as an instrument of genetic stability has already been referred to. Weaver & Cavers (1979) have drawn attention to the complexities which affect the level of the seed bank using as study material two

species of *Rumex*. They noted that there is evidence that conditions which promote seed dormancy also promote seed longevity in the soil, and examined the effect on the levels of the seed bank of conditions such as darkness and constant temperatures which tend to increase levels of dormancy following burial of freshly shed seed, and conditions such as germination and seed loss which deplete the bank. Their results suggest that in areas where the patterns of disturbance is variable the dynamics of the seed bank will be complex. Locations in which seeds are buried rapidly but then remain undisturbed are likely to contain large numbers of seeds; but in situations where seeds remain very close to the surface depletion would be rapid. Cohen (1966) has modelled the parameters affecting the significance and size of the soil-borne seed population in relation to its success as a strategy contributing to the survival of the population. He has proposed that the mechanics of the strategy depends on two basic factors - the rate of decay, or loss by predation etc, of seeds within the soil; and the reproductive success of germinated seeds. In those situations where prospects of reproductive success are high the optimal strategy would tend towards high annual germination rates. In those where prospects are low, low germination rates would be more effective - provided rates of decay or loss were not excessive. The situation is highly sensitive to modification by the local environment and interactions with the reproductive strategy of different species suggest that 'trimming' of the proportion of dormant seeds within populations, could be a highly effective means of responding to changing environments. Reports on the species constitution of the seed bank in relation to the occurrence of species in established vegetational complexes by Whipple (1978), Graber & Thompson (1978), Moore & Wein (1977) and Cheke et al. (1979), all draw attention to the oft-reported point that correspondence between the two is unusual. Evidently there are very wide differences in the extent to which different species depend on the seed bank as a part of their reproductive strategy, and it appears as an element which, rather notably, may play a major or negligible role depending largely on the species involved.

Seed production within a species is dependent on the proportion of total resources available for allocation to the construction of the seed and the storage reserves within it, and these resources are inevitably limited at levels defined by the plant's environment. Since there is often a balance of advantage between the competitive capability of large seeds and the prospects for effective dispersal related to numerous smaller seeds, crude increases in numbers at the expense of size, or vice versa, may fail to materially affect the reproductive capacity of the species decisively. This results only from an allocation of greater resources to seed production. Bostock & Benton (1979) have examined the allocation of resources available both for seed and vegetative reproductive strategies in five perennial species of Compositae and have discussed the effective contribution each strategy may make to survival and maintenance of the populations within a particular area. They make the point that selection does not act necessarily in the same direction in both seeds and vegetative reproduction, and that this may provide for both r-selected or predominantly K-selected strategies to proceed concurrently or consecutively in the same species at different intensities. Harmer & Lee (1978) have, similarly, compared vegetative and seed-based strategies in the viviparous grass *Festuca vivipara* and the closely related *F. ovina*, in relation to their occurrence in arctic-alpine areas of relatively high precipitation. They draw attention to the much greater resources contained within plantlets of *F. vivipara* compared to seeds of *F. ovina* and relate this to aspects of establishment, and even to improved soil stabilisation by roots developing profusely from plantlets in the autumn under cool moist conditions. In such locations, in which stress may be a dominant feature of the local environment resource, allocation may be clearly directed towards establishment rather than in seeking out new sites, by favouring or relying exclusively on vegetative forms of reproduction. McNamara & Quinn (1977) have extended such comparisons in a study of *Amphicarpum purshii*, a species of grass which produces both subterranean cleistogamous and aerial chasmogamous spikelets. It occurs as a pioneer species in secondary succession within which it maintains a strategy of producing a relatively large subterranean propagule with a high chance of establishment resulting from the natural advantages of its placement and seedling vigour, but remains the advantages of a smaller aerial propagule produced in large numbers with potentially greater genetic variability, and capacity for dispersal.

The manner in which local conditions govern the balance of advantage between strategies based on seed establishment or dispersal, have not been widely reported. McWilliams et al. (1968) report a positive regression between seed weight and latitude in *Amaranthus reflexus*, but confidence in the result is somewhat reduced by its dependance on a small number of populations with a northerly distribution. Schimpf (1977) has suggested that moisture availability may be of greater importance than length of growing season (broadly latitude) as a selective force affecting seed weight. He suggests that large seeds may be correlated with drier environments - a response also reported for *Pinus caribea* by Kandya et al. (1978). Janzen (1977a, b), has reported that the seeds produced by two Central American species of arboreal Leguminoseae, *Cassia grandis* and *Mucuna andreana* vary naturally from tree to tree and within the progeny of a single tree over at least a threefold fresh weight range, and has related this natural tendency to variations in the patterns of dispersal. *M. andreana* is wind-dispersed, *C. grandis* is deposited in the faeces of animals which consume the fruits, but in both cases Janzen suggests that a variety of seed sizes may play a part in producing an evenly dispersed seed shadow effect over a relatively wide area, which could be achieved less successfully if all seeds were of the same size.

Constraints which affect the balance of advantage held by smaller or larger seeds must clearly take into account a great variety of different elements in the reproductive strategy of the species, and several reports have recently appeared linking seed size and numbers with patterns of predation. Thus Bradford & Smith (1977) in a study of populations of two *Scheelea* species from Central America, found that most fruits were single seeded, but a few contained two or three seeds. The seeds from multiseeded fruits were smaller and contained fewer storage reserves than those from single seeded fruits, but since it was unusual for bruchid beetles to attack more than one seed in a fruit, they also displayed higher survival rates to seed maturity - conferring some overall advantage in spite of the smaller seed size. Other observations on the effects of predation on seed size stem from reports of the effects of bruchid beetles on seeds of *Cercidium floridum* by Mitchell (1977), and of birds on the fruits of *Stemmadenia donnell-smithii* by McDiarmid et al. (1977). The latter study attempts to assess the ways in which resources are allocated between different parts of the fruits such as the husk, the seed and the fleshy aril, in relation to prospects of successful dispersal and seedling establishment.

The significance of seed-based and vegetative reproductive strategies have also been examined by Keeley (1977) in two pairs of congeneric shrubs characteristic of the Californian chaparral. In each pair, one species depended entirely on seeds for regeneration after fire the other on seeds as well as subterranean burls. The study identified separate strategies for each of the four species examined, and there was no evidence, even under these rather closely defined conditions, of the attainment of a single reproductive strategy which closely fitted the nature of the conditions and was possessed by all four species. Instead it was clear, as has been found many times before, that each species could be characterized by its own solution to the problems set by the environment which related to its own particular combination of reproductive capacity, phenology and competitive capacity.

Phenotypic plasticity of germination responses in relation to geographical distribution

Variations in temperature responses over well-defined ranges; variations in the level of post-harvest dormancy, and rates of after-ripening; variations in degree of dependance on light and dark, sometimes accompanied by morphological variations in seed sizes, colour or shape (e.g. Edgar & Springfield, 1977) - these are examples of variability between individuals within a population. There is evidence that a proportion, sometimes a major proportion, of the expressed effects of these variations arise from genotype × environment interactions - particularly in response to ambient conditions experienced during the last few days of seed maturation on the parent plant, as reported for *Sesbania bispinosa* (Sharma et al., 1978) or in response to conditions experienced after dispersal of the seed under natural conditions, or in store. Phenotypic plasticity of this kind arising from variations in conditions experienced by the parent plants have been reported by Gutterman (1973), and changes developing after harvest are well known, widely reported and referred to collectively as after-ripening processes.

It appears that polymorphisms in germination responses are universally found in seeds in response to some combination of ambient conditions. A characteristic example of the phenomenon is contained in a report on the germination of *Hordeum jubatum* by Banting (1979) which notes that: 'Germination was greater at alternating than at constant temperature, and was increased more by alternating temperatures of 20 and 30 °C for 18 and 6 h, than by 10 and 15 °C for 10 and 14 h, respectively, in darkness. Germination was reduced by light, the effect being most pronounced when the light was continuous and at constant temperature'. It is characteristic of such results that they provide a pattern of almost continuous quantative variation in which any modification to the ambient conditions experienced by a set of seeds may result in more or less germination amongst samples of seeds set up for tests. Effects of this kind trim germination strategies in ways which may remain within particular recognisable qualitative bounds i.e. generalised requirements for chilling, light and darkness, high temperatures etc. - but which may produce quantitative variations at certain temperatures, or situations, which are highly significant in relation to the natural environment and ecology of the population or species involved. They may act to modify germination responses very effectively in ways which do not depend upon variations in genotype, but which can nevertheless affect different populations to different degrees or in different ways according to the geographical location and climatic characteristics of their origin as an inevitable consequence of interactions between environment and genotype. Very straightforward examples of this situation are represented by the seeds of species such as *Aegilops* (Datta et al., 1970) and *Ononis sicula,* and *Portulaca oleracea* (Gutterman, 1973) in which the level of dormancy at harvest time has been shown to be dependant upon the position of the individual seed or fruit within the inflorescence or the conditions experienced by the mother plant at the time seeds matured. More subtle effects are now being repeatedly found; for example Baskin & Baskin (1978b) have described the ways in which seeds of *Cyperus inflexus* after-ripen in the field during winter so that a proportion germinate the following spring. The ungerminated residue experience induced dormancy due to high temperatures during the summer, but respond to low temperature conditioning treatments through the winter which enables a further proportion to germinate the following spring. Repeated cycles of this sort may continue for several years resulting in successive flushes of seedlings in consecutive springs and summers. In a rather different way seeds of *Lactuca serriola* (Marks & Prince, 1979) undergo changes in their internal physiology as the seasons progress, which vernalise or devernalise the mechanisms which lead to the production of flower initials and stem elongation in the young plants following germination. The population metamorphoses successively between a largely vernalised or largely unvernalised condition depending on the season of the year and the temperature conditions experienced by exposed seeds. Responses of the seeds of a number of woodland species including *Silene dioica, Hyacinthoides non-scripta* and *Milium effusum* reported by Thompson (1975, 1980) and Thompson & Cox (1978) display levels of dormancy which are strongly affected by temperature before and after seed maturation in ways which suggest that higher proportions of seeds are capable of germination soon after harvest in souhern parts of the range of the species concerned, in sunny sheltered situations, or from individuals in which a high proportion of seeds become mature during warm spells in summer rather than in cooler conditions during the autumn. A germination model has been proposed for species under these conditions which suggests that genotype × environment interactions are sufficient to account for the great majority of population differences in germination behaviour, including in particular the polymorphic responses which divide populations into sets germinating at different seasons of the year.

Maintenance of polymorphic germination responses

The existence of germination polymorphisms and their long-term maintenance, without selective change in the direction of one or another of the particular responses involved, is a major feature characteristic of the germination process. There is little or no experimental evidence to show how such variations can be continuously maintained, although it is evident at least in a prima facie sense that its maintenance is paradoxically an essential part of the adaptive responses of many species - particularly those from areas with unpredictable, erratic or complex climatic patterns similar to those occurring in the oceanic regions of western Eurasia. It is now evident that many polymorphisms depend for their existence on phenotypic plasticity, and in these cases selective affects on genotype would not be expected. Nevertheless, there are many other situations in which differences in levels of germination under particular conditions appear in ways which appear independent of phenotypic variation.
The results of germination tests are habitually expressed in terms of percentage responses. They normally relate to the number of seeds which germinated under the influence of a stated set of conditions, and the result is frequently interpreted as though it measures a fractional proportion of some hypothetical whole - in other words the responses of the individuals which happened to germinate are used to measure the germinability of the set as a whole and this interpretation may be carried on to make comparisons between responses of different batches of seeds or between sets of the same batch under a variety of conditions. However, because of the way that tests are set up and scored the results obtained show no more than that under certain conditions a particular number of seeds, which it is assumed possess certain qualities in common, produced seedlings and a certain number, which it is assumed are in some way(s) different, remained ungerminated. The results of such tests are frequently interpreted in ways which strain the information they actually contain - especially when comparisons are made between different seed lots or responses to different conditions.
An example from experiments by Baskin & Baskin (1979) on the germination of *Aster pilosus* provides typical results of the diversity of germination responses which illustrate this problem. Numerous experimental regimes were set up and a wide variety of percentage germination rates obtained, of which the following are a small sample.

Treatment A: Seeds tested in light at 20/10 °C → 68 % germination
Treatment B: Seeds tested in light at 20 °C → 28 % germination
Treatment C: Seeds tested in dark at 20/10 °C → 7 % germination
Treatment D: Seeds tested in dark after stratification → 41 % germination

These results can be expressed with fiducial limits to indicate whether they represent significant variations, but they tell us nothing about the composition of the various fractions of the whole population relative to one another. For example there is no way of knowing whether all the seeds capable of germinating in conditions B and C were able also to germinate in response to condition A. It is not possible to suggest which individuals germinating in condition D would also have been capable of germinating in A etc. In other words there is no way of logically partitioning those individuals which do or do not germinate in different tests. It is usually taken for granted that those individuals which do germinate are in some way different from those which do not, but the only basis for this assumption is their behaviour in the test. If, in some way which is technically not possible, the test could be repeated with the same individuals packed back into their seed coats it might happen that the same numerical responses would recur (within statistical limits), but that the individuals contributing to the result would differ. In such a case it would be seen that the test measures the probability that a particular seed would germinate or remain dormant, and the result is made up from a number of individuals selected at random from the whole set according to probabilities defined by the condition of the seeds and their past experience.

In essence the problem is to determine whether the results obtained in germination tests depend on actual differences, genetic or phenotypic, between individual seeds, or whether they represent the probability that a seed of a particular composition will germinate given a particular set of conditions and, therefore, imply no genetic or phenotypic difference between individuals that do or do not germinate.

It is possible that the answer may lie in the direction of strict genetic control of each and every response, and that in the results just quoted for *Aster pilosus* the seven seeds germinating in darkness (condition C) were each in some way different to most, if not all the twenty-eight seeds germinating in light and constant temperature (condition B). A problem with this approach is that as test conditions and test responses vary so do the results, and, given an almost endless variety of responses, there appears to be little alternative but to quote almost endless genotypes to account for them. Secondly, such a system would appear most likely to be highly adaptive and unlikely to possess the property of maintaining continuous variation over time.

The alternative would propose a uniform genotype - in an ideal statement of the situation - in which the test simply measures the probability of a response leading to germination. Situations in which this kind of response occur have been reported. For example Harper & Benton (1966) examined responses of seeds of *Lepidium sativum, Clarkia elegans* and *Agrostemma githago* set out on the surface of a substrate which supplied water in a controlled and effectively limited fashion. They found that even in a 'homogenous' environment the numbers of seeds germinating was determined by the individual properties of each, how it lay, what its shape was, what contact it made with its substrate, and the evaporative loss at the seed coat/atmosphere interface. In this experiment the germination response

measured the probability that the water uptake balance would be sufficient to provide for germination, and because of the rather obvious physical or spatial or conformational factors involved, the possibility of genetic differences being responsible for observed differences in germination hardly arose.
Circumstances in which external conditions regulate germination independently of the genetic constitution of the seed in this and similar ways, may appear to be fundamentally different from the internal conditions which regulate responses to temperature, light conditioning treatments etc. But there is little evidence available to indicate the nature of such mechanisms. If they were found to depend on relatively simple rate limiting responses leading to the sequential production of particular products which promote or inhibit germination, or on changes in membrane permeability of limited extent, or other physical changes depending on integrated, mutually dependent sequences of change, there could be some presumption for direct genetic control involving mechanisms of relative simplicity. If, however, germination depends on large numbers of reactions proceeding independently, similar to those proposed recently by Labouriau (1978, 1979), which result in germination only when they arrive at some form of synchronisation or favourable configuration, the situation could well resemble more closely a model similar to the conditions provided by the physical constraints of Harper and Benton's experiment quoted earlier, and germination could then depend on the probability that reactions would 'line up' in favourable configurations.
Lack of evidence prevents further speculation, but fundamental to an understanding of the way that germination responses play a part, whether adapted or adaptive, in fitting populations of species for survival in particular locations is some information, now almost entirely lacking, on the genetic control of germination processes, and on the controlling mechanisms which produce the phenotypic plasticity which causes one seed to be dormant and another to germinate.

Acknowledgements

This review marks the end of my involvement in research in the botanical sciences and is the last contribution I shall attempt towards a better understanding of the subject. I am most happy to be able to take the opportunity to express my warmest thanks to those who have provided assistance both in my research work, and in the preparation of the results for publication.
In particular I should like to express my appreciation to:- Mr David Fox without whom I should never have had the equipment to do my experiments; to Mrs Jane Pescheira whose skill in draughtsmanship has provided illustrations for my publications; to Mrs Dorothy Williams and Mrs Pamela Bloomfield who have successfully deciphered my manuscripts to produce impeccable typescripts, and to all other members of the staff at the Royal Botanic Gardens, Wakehurst Place, whose help has supported, cheered, encouraged and enlightened me.

References

Archibold, O.W., 1979. Buried viable propagules as a factor in post-fire regeneration in northern Saskatchewan. Canadian Journal of Botany 57: 54-58.
Ashton, D.H., 1979. Seed harvesting by ants in forests of Eucalyptus regnans F. Muell. in central Victoria. Australian Journal of Ecology 4: 265-277.
Augspurger, C.K., 1979. Irregular rain cues and the germination and seedling survival of a Panamanian shrub (Hybanthus prunifolius). Oecologia 44: 53-59.
Baker, H.G., 1965. Characteristics and modes of origin of weeds. In: H.G. Baker and G. Ledyard Stebbins (Eds), The genetics of colonising species. Academic Press, New York and London.
Banting, J.D., 1979. Germination, emergence and persistence of foxtail barley. Canadian Journal of Plant Science 59: 35-41.
Bartolome, J.W., 1979. Germination and seedling establishment in California annual grassland. Journal of Ecology 67: 273-281.
Baskin, J.M. & C.C. Baskin, 1977a. Germination ecology of Sedum pulchellum Michx. (Crassulaceae). American Journal of Botany 64: 1242-1247.
Baskin, J.M. & C.C. Baskin, 1977b. Germination of common milkweed (Asclepias syriaca L.) seeds. Bulletin of the Torrey Botanical Club 104: 167-170.
Baskin, J.M. & C.C. Baskin, 1978a. A contribution to the germination ecology of Rumex crispus L. Bulletin of the Torrey Botanical Club 105: 278-281.
Baskin, J.M. & C.C. Baskin, 1978b. Seasonal changes in the germination responses of Cyperus inflexus seeds to temperature and their ecological significance. Botanical Gazette 2: 231-235.
Baskin, J.M. & C.C. Baskin, 1978c. The seed bank in a population of an endemic plant species and its ecological significance. Biological Conservation, 14, 125-130.
Baskin, J.M. & C.C. Baskin, 1979. The germination strategy of old-field aster (Aster pilosus). American Journal of Botany 66: 1-5.
Black, M., 1969. Light-controlled germination of seeds. In: H.W. Woolhouse (Ed.), Dormancy and Survival. Symposia of the Society for Experimental Biology 23: 193-217.
Blom, C.W.P.M., 1978. Germination, seedling emergence and establishment of some Plantago species under laboratory and field conditions. Acta Botanica Neerlandica 27: 257-171.
Bostock, C.J. & R.A. Benton, 1979. The reproductive strategies of five perennial compositae. Journal of Ecology 67: 91-107.
Bouchard, R.P. & E.H. Franz, 1977. Seed germination and habitat selection in Talinum teretifolium Pursh. Georgia Journal of Science 35: 159-169.
Bradford, D.F. & C.C. Smith, 1977. Seed predation and seed number in Scheelea palm fruits. Ecology 58: 667-673.
Bradshaw, A.D., 1965. Evolutionary significance of phenotypic plasticity in plants. Advances in Genetics 13: 115-155.
Cabiaux, C. & F. Devillez, 1977. Etude de l'influence des facteurs du milieu sur la germination et la levée des plantules du bouleau pubescent. Bulletin de la Societé Royale de Botanique de

Belgique 110(1-2): 96-112.
Charlton, J.F.L., 1977. Establishment of pasture legumes in North Island hill country. I. Buried seed population. New Zealand Journal of Experimental Agriculture 5(3): 211-214.
Cheke, A.S., W. Nanakorn & C. Yankoses, 1979. Dormancy and dispersal of seeds of secondary forest species under the canopy of a primary tropical rain forest in Northern Thailand. Biotropica 11(2):88-95.
Cohen, D., 1966. Optimising reproduction in a randomly varying environment. Journal of Theoretical Biology 12: 119-129.
Cole, N.H.A., 1977. Effect of light, temperature, and flooding on seed germination of the neotropical Panicum laxum Sw. Biotropica 9: 191-194.
Datta, S.C., M. Evenari & Y. Gutterman, 1970. The heteroblasty of Aegilops ovata L. Israel Journal of Botany 19: 463-483.
Davidson, D.W., 1977. Species diversity and community organization in desert seed-eating ants. Ecology 58: 711-724.
Davidson, D.W., 1977. Foraging ecology and community organization in desert seed-eating ants. Ecology 58: 725-737.
Davidson, E.D. & M.G. Barbour, 1977. Germination, establishment and demography of coastal bush lupine (Lupinus arboreus) at Bedega Head, California. Ecology 58: 592-600.
DeAngelis, D.L., C.C. Travis & W.M. Post, 1979. Persistence and stability of seed-dispersed species in a patchy environment. Theoretical Population Biology 16: 107-125.
Del Tredici, P., 1977. The buried seeds of Comptonia peregrina, The Sweet fern. Bulletin of the Torrey Botanical Club 104: 270-275.
Edgar, R.L. & H.W. Springfield, 1977. Germination characteristics of broadscale: A possible saline-alkaline site stabilizer. Journal of Range Management 30: 296-298.
Evans, R.A., B.L. Kay & J.A. Young, 1976. Influence of date of planting on emergence of cultivars of Trifolium hirtum All. and T. subterraneum L. Journal of Range Management 29: 333-337.
Farmer jr., R.E., 1978. Propagation of a southern Appalachian population of fringed gentian. Bulletin of the Torrey Botanical Club 105: 139-142.
Fenner, M., 1980a. The induction of a light requirement in Bidens pilosa seeds by leaf canopy shade. New Phytologist 84: 103-106.
Fenner, M., 1980b. The inhibition of germination of Bidens pilosa seeds by leaf canopy shade in some natural vegetation types. New Phytologist 84: 95-101.
Floyd, A.G., 1976. Effect of burning on regeneration from seeds in wet sclerophyll forest. Australian Forestry 39: 210-220.
Frankton, C. & I.J. Bassett, 1968. The genus Atriplex (Chenopodiaceae) in Canada, Part 1. Three introduced species: A. heterosperma, A. oblongifolia, and A. hortensis. Canadian Journal of Botany 46: 1309-1313.
Gill, A.M., 1976. Fire and the opening of Banksia ornata F. Muell. follicles. Australian Journal of Botany 24: 329-335.
Gorski, T., K. Gorska & J. Nowicki, 1977. Germination of seeds of various herbaceous species under leaf canopy. Flora 166: 249-259.
Gorski, T., K. Gorska & J. Rybicki, 1978. Studies on the germination of seeds under leaf canopy. Flora 167: 289-299.
Graber, R.E. & D.F. Thompson, 1978. Seeds in the organic layers and

soil of four beech-birch-maple stands. Forest Service Research Paper, 8 p.
Gramshaw, D. & W.R. Stern, 1977. Survival of annual ryegrass (Lolium rigidum Gaud) in a Mediterranean type environment. Effect of summer grazing by sheep on seed numbers and seed germination in autumn. II Persistence of viable seed. Australian Journal of Agricultural Research 28: 81-101.
Grubb, P.J., 1977. The maintenance of species-richness in plant communities: the importance of the regeneration niche. Biological Reviews 52: 107-145.
Gutterman, Y., 1973. Differences in the progeny due to day length and hormone treatment of the mother plant. In: W. Heydecker (Ed.), Seed Ecology, Butterworths, London, p. 59-98.
Handel, S.N., 1978. The competitive relationship of three woodland sedges and its bearing on the evolution of ant-dispersal of Carex pedunculata. Evolution 32: 151-163.
Hare, J.D. & D.J. Futuyma, 1978. Different effects of variation in Xanthium strumarium L. (Compositae) on two insect seed predators. Oecologia 37: 109-120.
Harmer, R. & J.A. Lee, 1978. The germination and viability of Festuca vivipara (L.) Sm. plantlets. New Phytologist 81: 745-751.
Harper, L.J., 1977. Population biology of plants. Academic Press, London, New York and San Francisco.
Harper, L.J. & R.A. Benton, 1966. The behaviour of seeds in soil. the germination of seeds on the surface of a water supplying substrate. Journal of Ecology 54: 151-166.
Harper, J.L., P.H. Lovell & K.G. Moore, 1970. The shapes and sizes of seeds. Annual Review of Ecology and Systematics 1: 327-356.
Harper, J.L., J.T. Williams & G.R. Sagar, 1965. The behaviour of seeds in soil. I. The heterogeneity of soil surfaces and its role in determining the establishment of plants from seed. Journal of Ecology 53: 273-286.
Helliwell, D.R. & A.F. Harrison, 1978. Variation in the growth of different seeds of Acer pseudoplatanus and Betula verrucosa grown on different soils. Forestry 51: 37-46.
Hnatiuk, S.H., 1979. A survey of germination of seeds from some vascular plants found on Aldabra Atoll. Journal of Biology, Osaka City University 6: 105-114.
Hodgkinson, K.C., G.H. Harrington & G.M. Miles, 1980. Composition, spatial and temporal variability of the soil seed pool in a Eucalyptus populnea shrub woodland in central New South Wales. Australian Journal of Ecology 5: 23-29.
Howe, H.F., 1977. Bird activity and seed dispersal of a tropical wet forest tree. Ecology 58: 539-550.
Hurka, H. & M. Benneweg, 1979. Patterns of seed size variation in populations of the common weed Capsella bursa-pastoris (Brassicaceae). Biologisches Zentralblatt 98: 699-709.
Janzen, D.H., G.A. Miller, J. Hackforth-Jones, C.M. Pond, K. Hooper & D.P. JANOS, 1976. Two Cost Rican bat-generated seed shadows on Andira intermis (Leguminosae). Ecology 57: 1068-1075.
Janzen, D.H., 1977a. Variation in seed weight in Costa Rican Cassia grandis (Leguminosae). Tropical Ecology 18: 177-186.
Janzen, D.H., 1977b. Variation in seed size within a crop of a Costa

Rican Mucuna andreana (Leguminosae). American Journal of Botany 65: 347-349.
Janzen, D.H., 1977c. Intensity of predation of Pithecellobium saman (Leguminosae) seeds by Merobruchus columbinus and Stator limbatus (Bruchidae) in Costa Rican deciduous forest. Tropical Ecology 18: 162-176.
Johns, G.G. & L.R. Greenup, 1976a. Pasture seed theft by ants in northern New South Wales. Australian Journal of Experimental Agriculture and Animal Husbandry 16: 249-256.
Johns, G.G. & L.R. Greenup, 1976b. Predictions of likely theft by ants of oversown seed for the Northern Tablelands of New South Wales. Australian Journal of Experimental Agriculture and Animal Husbandry 16: 257-264.
Jones, R.M. & T.R. Evans, 1977. Soil seed levels of Lotononis bainesii, Desmodium intortum and Trifolium repens in subtropical pastures. Journal of the Australian Institute of Agricultural Science 43: 164-166.
Jouret, M.F., 1976. Relationships between origin and germinative behaviour in seeds of Impatiens balfourii. Bulletin de la Société Royale de Botanique de Belgique 109: 209-212.
Jouret, M.F., 1977. Relation entre la dormance seminale et la chorologie de diverses especes du genre Impatiens L. Bulletin de la Société Royale de Botanique de Belgique 110: 119-128.
Kandya, A.K., C.R. Venator & E.F. Brunif, 1978. Seed and seedling characteristics in Pinus caribaea Mor. as geographical variation parameters: a view point. Tropical Ecology 19: 65-69.
Keeley, J.R., 1977. Seed production, seed populations in soil, and seedling production after fire for two congeneric pairs of sprouting and nonsprouting chaparral shrubs. Ecology 58: 820-829.
Kirkpatrick, B.L. & F.A. Bazzaz, 1979. Influence of certain fungi on seed germination and seedling survival of four colonizing annuals. Journal of Applied Ecology 16: 515-527.
Labouriau, L.G., 1978. Seed germination as a thermobiological problem. Radiation and Environmental Biophysics 15: 345-366.
Labouriau, L.G., 1979. Isothermal germination rates in seed of Dolichos biflorus L. Boletin Socidad Venezolana Naturales 34: 73-112.
Levins, R., 1968. 'Evolution in a changing environment'. Princetown University Press, Princetown, N.J., USA.
Lewis, H., 1962. Catatrophic selection as a factor in speciation. Evolution 16: 257-271.
McDiarmid, R.W., R.E. Ricklefs & M.S. Foster, 1977. Dispersal of Stemmadenia donnell-smithii (Apocynaceae) by birds. Biotropica 9: 9-25.
McNamara, J. & J.A. Quinn, 1977. Resource allocation and reproduction in populations of Amphicarpum purshi (Gramineae). American Journal of Botany 64: 17-23.
McNaughton, S.J., 1966. Ecotype function in the Typha community-type. Ecological Monographs 36: 297-325.
McWilliams, E.L., R.Q. Landers & J.P. Mahlstede, 1968. Variation in seed weight and germination in populations of Amaranthus retroflexus L. Ecology 49: 290-296.
Makhnev, A.K. & O.V. Makhneva, 1979. Patterns of intraspecific

variability of birch for biological seed characters in the Northern Urals. Soviet Journal of Ecology 10: 105.113.

Marks, M.K. & S.D. Prince, 1979. Induction of flowering in wild lettuce (Lactuca serriola L.) II Devernalization. New Phytologist 82: 357-363.

Martell, A.M., 1979. Selection of conifer seeds by deer mice and red-backed voles. Canadian Journal of Forest Research 9: 201-204.

Mitchell, R., 1977. Bruchid beetles and seed packaging by palo verde. Ecology 58: 644-651.

Moore, J.M. & R.W. Wein, 1977. Viable seed populations by soil depth and potential site recolonization after disturbance. Canadian Journal of Botany 55: 2408-2412.

Morley, F.H.W., 1958. The inheritance and ecological significance of seed dormancy in subterranean clover (Trifolium subterraneum L.). Australian Journal of Biological Sciences 11: 261-274.

Morley, F.H.W. & J. Katznelson, 1965. Colonization in Australia by Trifolium subterraneum L. In: H.G. Baker and G. Ledyard Stebbins (Eds), The Genetics of Colonizing species. Academic Press, New York and London.

Mott, J.J., 1978. Dormancy and germination in five native grass species from savannah woodland communities of the Northern Territory. Australian Journal of Botany 26: 621-631.

Napp-Zinn, K., 1976. Population, genetical and gene geographical aspects of germination and flowering in Arabidopsis thaliana. Arabidopics Information Service 13: 30-33.

New, J.K., 1958. A population study of Spergula arvensis. I. Two clines and their significance. Annals of Botany 22: 457-477.

New, J.K., 1959. A population study of Spergula arvensis. II. Genetics and breeding behaviour. Annals of Botany 23: 23-33.

Noble, J.C. & R.D.B. Whalley, 1978. The biology and autecology of Nitraria L. in Australia. II. Seed germination, seedling establishment and response to salinity. Australian Journal of Ecology 3: 165-177.

Ødum, S., 1965. Germination of ancient seeds; floristical observation and experiments with archaeologically dated soil samples. Dansk Botanisk Arkiv 24: 1-70.

Oesau, A., 1975. Untersuchungen zur Keimung und Entwicklung des Wurzelsystems in der Gattung Melampyrum L. (Scrophulariaceae). Beitrage zur Biologie der Pflanzen 51: 121-147.

Oomes, M.J.M. & W.Th. Elberse, 1976. Germination of six grasslands herbs in microsites with different water contents. Journal of Ecology 64: 745-755.

Panetta, F.D., 1979. Germination and seed survival in the woody weed, groundsel bush (Baccharis halimifolia L.). Australian Journal of Agricultural Research 30: 1067-1077.

Purdie, R.W., 1977. Early stages of regeneration after burning in dry sclerophyll vegetation. II. Regeneration by seed germination. Australian Journal of Botany 25: 35-46.

Raguse, C.A., 1977. Germination of California annual range plants in response to a summer rain. Agronomy Journal 69: 327-328.

Reichman, O.J. & D. Oberstein, 1977. Selection of seed distribution types by Dipodomys merriami and Perognathus amplus. Ecology 58: 636-643.

Salisbury, E.J., 1942. The reproductive capacity of plants. G. Bell and Sons Ltd., London.
Schimpf, D.J., 1977. Seed weight of Amaranthus retroflexus in relation to moisture and length of growing season. Ecology 58: 450-453.
Semenza, R.J., J.A. Young & R.A. Evans, 1979. Influence of light and temperature on the germination and seedbed ecology of common mullein (Verbascum thapsus). Weed Science 26: 577-581.
Sharma, K.P. & B. Gopal, 1978. Seed germination and occurrence of seedlings of Typha species in nature. Aquatic Botany 4: 353-358.
Sharma, M.M., N.K. Sharma & D.N. Sen, 1978. A new report on differential seed coat dormancy in Sesbania bispinosa (Jacq.) W.F. Wight. Folia Geobotanica and Phytotaxonomica 13: 95-98.
Siddiqi, M.Y., P.J. Myerscough & R.C. Carolin, 1977. Studies in the ecology of coastal heath in New South Wales. IV. Australian Journal of Ecology 1: 175-183.
Silvertown, J., 1980. Leaf-canopy-induced seed dormancy in a grassland flora. New Phytologist 85: 109-118.
Slade, E.A. & D.R. Causton, 1979. The germination of some woodland herbaceous species under laboratory conditions: a multifactorial study. New Phytologist 83: 549-557.
Sonia, L. & M.R. Heslehurst, 1978. Germination characteristics of some Banksia species. Australian Journal of Ecology 3: 179-186.
Stapanian, M.A. & C.C. Smith, 1978. A model for seed scatterhoarding: coevolution of fox squirrels and black walnuts. Ecology 59: 884-896.
Stearns, F. & J. Olson, 1958. Interaction of photoperiod and temperature affecting germination in Tsuga canadensis. American Journal of Botany 45: 53-58.
Stebbins, G.L., 1958. Longevity, habitat and release of genetic variability in the higher plants. Cold Spring Harbor Symposia on Quantitative Biology 23: 365-578.
Sullivan, T.P., 1978. Lack of caching of direct-seeded douglas fir seeds by deer mice. Canadian Journal of Zoology 56: 1214-1216.
Sullivan, T.P., 1979. The use of alternative foods to reduce conifer seed predation by the deer mouse, (Perimyscus maniculatus). Journal of Applied Ecology 16: 475-495.
Taylorson, R.B. & H.A. Borthwick, 1969. Light filtration by foliar canopies: significance for light-controlled weed seed germination. Weed Science 17: 48-51.
Thompson, K., 1979. The occurrence of buried viable seeds in relation to environmental gradients. Journal of Biogeography 5: 425-430.
Thompson, K. & J.P. Grime, 1979. Seasonal variation in the seed banks of herbaceous species in ten contrasting habitats. Journal of Ecology 67: 893-921.
Thompson, P.A., 1969. Effects of after-ripening and chilling treatments on the germination of species of Gentiana at different temperatures. Journal of Horticultural Science 44: 343-58.
Thompson, P.A., 1970a. Effects of temperature, chilling and treatment with gibberellins on the germination of Primula species. Journal of Horticultural Science 45: 175-185.
Thompson, P.A., 1970b. Germination of species of Caryophyllaceae in

relation to their geographical distribution in Europe. Annals of Botany 34: 427-449.
Thompson, P.A., 1973. Effects of cultivation on the germination character of the corncockle (Agrostemma githago L.). Annals of Botany 37: 133-154.
Thompson, P.A., 1975. Characterization of the germination responses of Silene dioica (L.) Clairv., populations from Europe. Annals of Botany 39: 1-19.
Thompson, P.A., 1980. Germination strategy of a woodland grass: Milium effusum L. Annals of Botany 46 (in press).
Thompson, P.A. & S.A. Cox, 1978. Germination of the bluebell (Hyacinthoides non-scripta (L.) Chouard) in relation to its distribution and habitat. Annals of Botany 42: 51-62.
Thompson, P.A., S.A. Cox & R.H. Sanderson, 1979. Characterization of the germination responses to temperature of lettuce (Lactuca sativa L.) achenes. Annals of Botany 43: 319-334.
Thompson, P.A. & D.J.C. Fox, 1976. The germination responses of vegetable seeds in relation to their history of cultivation by man. Scientia Horticulturae 4: 1-14.
Thompson, P.A., P. Newman & P.D. Keefe, 1979. Germination of species of Alstroemeria L. Gartenbauwissenschaft 44: 97-102.
Thurston, J.M., 1957. Morphological and physiological variation in wild oats (Avena fatua L. and A. ludoviciana Dur.) and in hybrids between wild and cultivated oats. Journal of Agricultural Science 49: 261-274.
Thurston, J.M., 1962. An international experiment on the effects of age and storage conditions on viability and dormancy of Avena fatua seeds. Weed Research 2: 122-129.
Ungar, I.A., 1979. Halophyte seed germination. Botanical Review 44: 233-264.
Ungar, I.A., 1979. Seed dimorphism in Salicornia europaea L. Botanical Gazette 140: 102-108.
Valk, A.G. van der & C.B. Davis, 1976. The seed banks of prairie glacial marshes. Canadian Journal of Botany 54: 1832-1838.
Vazquez-Yanes, C., 1980. Light quality and seed germination in Cecropia obtusifolia and Piper auritum from a tropical rain forest in Mexico. Phyton 38: 33-35.
Vegte, F.W. van der, 1978. Population differentiation and germination ecology in Stellaria media (L.) Vill. Oecologia 37: 231-245.
Waite, S. & M.J. Hutchings, 1978. The effects of sowing density, salinity and substrate upon the germination of seeds of Plantago coronopus L. New Phytologist 81: 341-348.
Waite, S. & M.J. Hutchings, 1979. A comparative study of establishments of Plantago coronopus L. sown randomly and in clumps. New Phytologist 82: 575-583.
Ware, S.A. & E.P. Quarterman, 1969. Seed germination in cedar glade Talinum. Ecology 43: 654-669.
Warr, H.J., D.R. Savory & A.K. Bal, 1979. Germination studies of bakeapple (Cloudberry) seeds. Canadian Journal of Plant Science 59: 69-74.
Watkinson, A.R., 1978. The demography of a sand dune annual Vulpia fasciculata. III. The dispersal of seeds. Journal of Ecology 66: 483-498.

Weaver, S.E. & P.B. Cavers, 1979. Dynamics of seed populations of Rumex crispus and Rumex obtusifolius (Polygonaceae) in disturbed and undisturbed soil. Journal of Applied Ecology 16: 909-917.

Went, F.W., 1955. The ecology of desert plants. Scientific Annals 192: 68-75.

Whipple, S.A., 1978. The relationship of buried, germinating seeds to vegetation in an old-growth Colorado subalpine forest. Canadian Journal of Botany 56: 1505-1509.

Whitcombe, J.R. & W.J. Whittington, 1972. The effects of selection for reduced dormancy in charlock (Sinapis arvensis). Heredity 29: 37-49.

Whitford, W.G., 1978. Foraging in seed-harvester ants (Pogonomyrmex spp. Ecology 59: 185-189.

Whittington, W.J., 1973. Genetic regulation of germination. In: W. Heydecker (Ed.), Seed Ecology. Butterworth, London, p. 5-30.

Williams, J.T. & J.L. Harper, 1965. Seed polymorphism and germination. I. The influence of nitrates and low temperatures on the germination of Chenopodium album. Weed Research 5: 141-150.

Yeaton, R.I., 1978. Some ecological aspects of reproduction in the genus Pinus L. Bulletin of the Torrey Botanical Club 105: 306-311.

Young, J.A. & R.A. Evans, 1979. Barbwire Russian thistle seed germination. Journal of Range Management 32: 390-394.

Young, J.A. & R.A. Evans, 1979. Arrowleaf balsamroot and mules ear seed germination. Journal of Range Management 32: 71-74.

Zhiryakov, V., 1977. Role of mammals as consumers of seeds of Schrenk's spruce. Soviet Journal of Ecology 8: 165-168.

Zlobin, Y.A., 1977. Ecology of germinating seeds of Scotch pine and Norway spruce. Soviet Journal of Ecology 8: 29-33.

Physiological and biochemical events in seed development

Daphne J. Osborne

During the development of a mature seed from a fertilised ovule several clearly defined tissues are differentiated. Some remain meristematic, others are secretory or function as storage loci for proteins and lipids, others serve a highly protective role for the enclosed embryo. These distinct tissue types become evident early in the progressive rounds of cell divisions that make up the mature seed. In certain tissues the cells remain alive and like those of the embryo, they are reactivated to full metabolic activity at germination, whilst in others the cells die during the organised desiccation that precedes the shedding and dispersal of the dry seed. In this review of many of the relevant papers on these topics published from 1974-1980, account has been taken of how each tissue type performs its specific function towards ensuring the survival of the new embryonic individual. Although no one tissue develops in isolation from its neighbours, nor is insulated from the influences that control the development of the whole seed in relation to the mother plant, it is none the less convenient to consider each of the tissues in turn before appraising the importance of the new findings for the understanding of the co-ordination of seed development.

Independence of the embryo

The anatomical pattern of differentiation from the single cell zygote of the fertilized ovule to the dry multi-cellular structures of the seed have been well documented for all the important crop plants and major weeds of the World though new features are constantly being noted or old features reappraised. As one example, a study of the appendages (the strophiole or rim-aril) on either side of the funicle in species of *Eriosema* has revealed that they contain storage lipid: this has led to the proposal that these arils are attractive to animals and therefore should aid in seed dispersal (Grear & Dengler, 1976).

Much effort has been spent over the past decade in attempts to interpret the control of differentiation pathways that lead to these anatomical distinctions and at a Symposium for Developmental Biology, McClintock (1978) drew attention to a basic concept in development. In considering the pattern of pigmented spots in the aleurones of kernels in hybrid maize mutants, she points out that these surface areas and their underlying cells could represent a single clone initiated by an activation of specific genetic changes very early in endosperm development, probably in one cell located within the embryo-sac complex. Such events occurring at the locus of a gene can regulate when and to what precise degree that gene should function during different stages of development, the molecular mechanism being operative without inducing permanent alterations of the DNA at the locus. At least we must believe the latter to be so if we subscribe to the concept of totipotency for all plant cells. The DNA complement of plant cells is not, however, immutable.

The introduction of foreign DNA with the ensuing permanent change of physiological and biochemical characters was long suspected following *Agrobacterium tumifaciens* infection of shoots. Cells of the typical bacteria-free crowngall tumours that then develop contain a fragment of a plasmid DNA as a permanent addition covalently linked to the plant's nuclear DNA (Yadav et al., 1980). The extra DNA first passes into the cells as a plasmid inclusion within the bacterium. Considering the many lines of evidence that led to these conclusions, it is not surprising that attempts are made to introduce foreign DNA into seeds to modify their genetic complement and hopefully improve the biochemical performance of the new plant. Korohoda & Strzalka (1979) describe experiments with maize in which DNA excised from a variety of *indurata* bearing dominant homozygous alleles of a *Su* gene (grains with a smooth surface) was supplied in the imbibition medium to grains of a variety of *sacharata* with *Su* recessive alleles (wrinkled grains). Some of the grains so treated produced plants with characteristics of the DNA donor or intermediate between donor and recipient. F1 progeny of modified grains produced grains or leaves with donor, recipient or intermediate characters. Analyses of protein, amino acids and sugar content indicated grains of certain types had higher contents than either original donor or recipient plants. The degrees of success achieved in transforming these crop plants with exogenous DNA were attributed to the initial vacuum treatment of grains with DNA solutions containing amounts of protamine equal to the DNA (w/w). The mechanisms involved here are not understood, but in the long term, these experiments indicate possibilities for transformation of crop plants which may prove to be of value for genetic manipulation.

Although, in the absence of external interference, the fertilized ovule has an inherent genetically controlled blue-print that dictates the eventual formation of characteristic root and shoot meristems, storage cells and enveloping layers, early phenotypic expression is also greatly under the control of maternal tissues. Abortion is a common phenomenon either through improper divisions of the zygote but more often from degeneration of the endosperm. Seed disfunction as high as 20 % is reported in *Abelmoschus esculentus* (Bhatnagar & Chandra, 1976) and 26 % in the commercially important pistachio nut (Bradley & Crane, 1975); unsuitable root-stocks appear to increase the numbers of abortive seeds in pistachio (Crane, 1975). In crossings between diploid and tetraploid *Trifolium* species (important to breeding programmes aimed at obtaining altitude-tolerant and virus-resistant strains of white clover) failure of the endosperm to form functional haustoria with maternal tissue resulted in starvation of the embryo and appeared to be a major cause of seed failure (Williams & White, 1976). A likely incompatability between hybrid and maternal tissues is implied, since some of the embryos that survive to the 'heart-shaped' stage seemed suitable for in vitro embryo culture. Certainly in vitro pollination of the silks of excised maize ovaries was successfully achieved in petri dishes of nutrient agar by Gegenbach (1977). Kernels were formed with or without endosperm with a 70 % success rate for embryo germination. That the material plant conveys no specific irreplaceable input after fertilization and is therefore theoretically superfluous for nursing of an embryo or for its viable seed formation is further emphasised by in vitro pollination of excised placental ovules of *Narcissus* (Balatková et al., 1977). Provided low temperature (15 °C) was maintained and 2 days post-anthesis placentas were used, glossy black viable seeds could be produced. Such culture systems are of obvious value to plant breeders, but a less rigorous technique has been demonstrated successfully for rye grass and orchard grass (Shimizu, 1978b). Ears harvested before flowering and kept in 3 % sucrose culture solution would produce some seeds. However, good seed yields with normal seed weight and germination performance were obtained when ears and attendant flag leaf, were cultured. Such results emphasise that whilst the mother plant is not essential for seed differentiation, successful culture techniques can suitably utilize the considerable advantages of additional parental nutrition.

Relationship of endosperm and embryo

Both the rate and the duration of dry matter accumulation before seed desiccation are of great importance to the potential germination success of the new seed and for the nutritive value of the grain. One problem that has plagued the rye-wheat Triticale hybrids is the large proportion of caryopses that shrivel at desiccation. In studies of normal and shrivel-prone lines, Salminen & Hill (1978) considered involvement of the parent plant (redirected photosynthetic activity, reduced translocation or early senescence), but concluded that the problem lay in low conversion to storage forms of translocated metabolites in seeds of shrivel-lines, and to accumulation of free sugars; the resulting high osmotic potentials and turgor pressures would then drive seed enlargement beyond the ultimate filling capacity. Such seeds might therefore shrivel at desiccation. Whether this be the mechanism for shrivelling or not remains to be established, but it could accord with earlier records of abnormal endosperm development (Bennett, 1974) and again implicates a control of seed formation by genetic information of the zygote, independent of the nutritional function of the maternal tissues.
Full understanding of factors regulating seed enlargement and those that cause the enlargement process to stop are clearly critical to a manipulation of plants to produce maximum yields of high quality seed (Bopp, 1979). Radley's (1978) experiments with wheat are therefore of considerable interest. They revealed that the number of both endosperm and aleurone cells is some 60 % greater in the giant third grains which develop when the two lowest grains of each spikelet are removed. Furthermore, during early development, incorporation of ^{14}C-sucrose by detached ears or halved grains is also greater in the giants than in the normal grains. She suggested, therefore, that the enhancement of starch synthesis and accumulation observed, may reflect a suppression of some factor which normally

prevents endosperm (and aleurone) cell numbers from reaching their maximum value. An over-production of storage tissue would, after all, have little evolutionary merit. That endosperm cell number or cell size could determine grain filling is not a new concept, but experiments designed to test storage capacity versus carbohydrate supply are relatively few. Manipulation of tillers and ears in wheat when endosperm cells were still dividing (Martinez-Carrasco & Thorne, 1979) showed that endosperm cell numbers were not altered by treatments that caused differences in grain dry weight. The factors regulating relative cell numbers of each tissue class in developing seeds still remains unresolved. (See later discussion on hormones and seed development.)

In sugarbeet (Jassem, 1973), intraspecific hybridisation of diploids, tetraploids and particularly triploids can lead to excessive physiological and anatomical disturbances in the development of the endosperm and this is one of the most common causes for decay in the seed: the degeneration of the embryo being a dependent secondary effect. In the intact plant the role of endosperm is certainly critical for the formation of a viable seed with good germination potential. Any abnormality or degeneration during seed development restricts proper embryo differentiation and failure of the cells to fulfil their proper storage role for carbohydrate, lipid or protein impairs that embryo's chances to establish as a successful seedling. The abnormalities in cell wall formation described by Sokolov & Kramarenko (1974) for *Iris* are therefore highly relevant. Normal endosperm at first undergoes a definite number of nuclear divisions to form a polynuclear cenocyte, cell walls are then laid down in the late telophase of the next and all subsequent division cycles. Abnormal endosperms appear to achieve the full number of divisions of the cenocyte stage and may even begin to deposit fragmoplasts or cell walls between the telophase nuclei. Varying degrees of wall formation are observed, but after several further rounds of nuclear division the cell walls or fragmoplasts are gradually destroyed and the free nuclear condition is again established. Clearly, this is not a simple failure of wall formation between nuclei but rather a process of deposition followed by active breakdown. In the event, any proper endosperm function as a storage tissue is precluded.

The specific biochemical performance of endosperm cells can sometimes be correlated with recognisable cytological differentiation of the nuclei. In *Cucurbita pepo* and *Bryonia dioica* (Turała-Szybowska, 1974) nuclei increase in size by endomitotic DNA replication to yield nuclei of varying ploidy, up to 384n or 48n respectively and show different endochromocentres and chromomere structures in different parts of the endosperm. A seasonal variation was also noted, so the factors determining the formation of specific nuclear structures are not clear. Nonetheless, the nutrition of the organism, the temperature during endosperm development and the level of polyploidy obtained were all considered to play an important role in eventual endosperm function. As Stebbins (1976) says in his thoughtful review of Seeds, Seedlings and the Origin of Angiosperms 'hypotheses about the origin and early evolution of angiosperms ... are based on distinctive characters affecting the physiology of seeds and seedlings. Most of the characteristics represent adapta-

tions for increasing the efficiency with which angiosperms exploit their environment. The early and rapid development of the endosperm even before that of the embryo is one such feature.' As always, there are extreme examples of the rule. In orchids and some herbaceous parasitic plants (Kumar, 1977), organ differentiation of a recognisable embryo occurs so late that it is an event of seed germination rather than of seed development. The cellular structure (or seed) that is shed is composed predominantly of starch containing endosperm cells, the embryo proper developing at the expense of the endosperm only after a filamentous outgrowth of the seed has made contact with a host root or when suitable levels of hormones or germination stimulants are attained in vivo or supplied in vitro.
At the other extreme, in the peanut (*Arachis hypogaea*) differentiation of the embryo in the mature seed has been reported to be so far advanced as to include flower primordia. A careful re-examination of seed from plants grown in different natural and controlled environments has however failed to confirm this (Staritsky, 1973).
In a developmental analysis of the growth of embryo and seed coats in four cultivars of a legume (*Phaseolus vulgaris*), Hsu (1979b) sought to determine what factors regulated the final differences in seed size. He concluded that there was a similar duration of development in all, but for large seed cultivars there was, at first, a faster growth rate of all parts and an earlier accumulation of storage materials in the embryos. In these in vitro studies, seed coats and embryos appeared to follow a similar and linked growth pattern. This contrasts with histological and chemical analyses of red currant seeds developing in vitro (Zatyko et al., 1979). Compared to seed coats of seeds maturing under natural conditions, those of cultured seeds consisted mainly of fleshy parenchymatous tissues reminiscent of the ripe fruit, though derived from a different original cell source. They call attention to the latent potentialities of the seed tissues to produce new organised forms when subjected to changed environmental or nutritional conditions.
There have been many chemical analyses of seeds during development, maturation and ripening. For cereals, see the careful work of Breidert & Schön (1979a and b) for summer barley. For legumes, see detailed results of Konno (1979) for soybean; Chatterjee et al. (1978) for pea. All show that the availability of precursors for macromolecule synthesis and deposition of polysacharides, protein or lipid are unlikely to be limiting seed development in vivo.
In vitro ovules with embryos of very few cells can be successfully reared and germinated, so informational inputs to normal differentiation must reside either totally within the cells of the zygote (as, for example, in an amphibian oocyte) or, partly in cells of the ovule in near juxtaposition to the zygote (as might be operative in the placental ovules of *Narcissus* (Balatkhová et al., 1977). In vivo, however, hormonal and nutritional signals are constantly passing to the embryo so that although a pea zygote will always differentiate a pea embryo, its size, metabolic activity and the general phenotypic expression and final vigour of the mature seed reflects the history of the association between the embryo and the mother plant. However, there is some evidence (Lucas & Milbourn, 1979) that seed nutrition may to some extent be influenced as early

as the pre-anthesis stage. Results of $^{14}CO_2$ feeding to leaves of *Phaseolus vulgaris* at or just before flowering suggest that the stem internode below the fed leaf plays a temporary storage role, supplying sugars for early pod fill.

Assimilate accumulation

Unravelling the contribution of different plant parts to overall seed production has not been an easy task though its importance, particularly in agriculture, is immense. In multi-fruited plants, analysis of seeds is confounded unless account is taken of the position of the pod or ear upon the stem and the distance of the fruit from the subtending photosynthetic supply organ. In following yields of seeds of two cultivars of *Vicia faba* - Abdel-Rahman & Youssef (1977) - or a similar paper - Youssef & Abdul-Rahman (1977) - showed that the upper position of the plant was less productive than the lower part. That competition with other pods is of particular importance in seed development is shown in experiments of Openshaw et al. (1979). On removing most pods on nodulating and non-nodulating lines of soybeans, higher weight and percentage protein in both lines was found, indicating that greater nitrogen fixation through nodulation was not the cause of the higher yield. A more detailed analysis of pod position and yield in hand-pollinated oil seed rape plants (Diepenbrock & Geisler, 1979) showed that seeds from axillary branches initially developed more rapidly than those on the main stems. However, hulls on the main branch retained a capacity to deposit storage lipid for longer and so outstripped their axillary branch competitors at seed maturity. The highest seed yield per pod was therefore greatest on the main branch. Furthermore, the proportion of the undesirable linolenic acid was low and the desirable oleic fatty acid relatively high. From the results, the authors describe the requirements for an ideal high-yielding rape plant. These include increasing the number of pods in the main branch, decreasing the number of lateral branches, encouraging a simultaneous development of all the pods and developing plants with hulls that actively function as photosynthetic and exporting tissues supplying the seeds (Murray, 1979). By feeding $^{14}CO_2$ to intact rape plants or to attached single leaves or siliques, Brar & Thies (1977) followed the accumulation of dry matter in the seeds. During the period of most active accumulation, relative photosynthetic activity of seeds, fruit walls (hulls) and stem parts bearing the siliques were respectively 2 %, 28 % and 41 % (leaves = 100 %) and the amount of ^{14}C transported from each organ to the seeds was 37 % from leaves, 32 % from the fruit walls and 31 % from the stems. These results reinforce the importance of fruit walls (hulls or siliques) as the source of assimilates and demonstrated that from their own photosynthetic efforts, seeds contributed no more than 1 % of the total ^{14}C which they accumulated. That rape seeds fix so little CO_2 for themselves is revealing for it was thought that the high level of linolenic acid (8-12 %) in the ripening cotyledons could be due to the well developed and photosynthetically active chloroplasts which they contain, the linolenic acid being a constituent of the chloroplast membranes. They further found that chlorophyll content, the number of stomata and the net photosynthetic activity of the seed and silique walls of a mutant were higher than in the parent cultivar despite lower linolenic acid contents in the seed wall. Studies of mutants could provide pointers indicating further ways that modelling of an ideal plant might proceed.

The functional value of the photosynthetic activity by fruit and seed is still somewhat hard to assess in the carbon economy of seed development and probably varies in response to different external circumstances. In attempts to obtain more precise information, the net gas exchange between detached seeds and developing pods was studied in *Lupinus albus* by Atkins & Flinn (1978). In the gas space in the pod, CO_2 content was normally higher than 1.0 % and from $^{14}CO_2$ labelling experiments at these concentrations it was clear that although CO_2 fixation occurred in the seed there was a net CO_2 efflux and 95 % of the $^{14}CO_2$ was fixed by the pod. The fact that young lupin seeds were rich in phosphoenolpyruvate (PEP) carboxylase and low in chlorophyll and ribulosediphosphate (RUDP) carboxylase could account for the initial high (about 60 %) and rapid (1 min) labelling of malate compared with phosphoglyceric acid (about 10 %). Possibly, therefore, PEP carboxylase functions in recycling the seed respiratory CO_2 by dark fixation from the pod gas space.

The presence of stomata in fruit walls and on seed coats is relatively common and in *Eschscholzia*, where only those on the outer fruit wall have the capacity to open and shut, Jernstedt & Clark (1979) believe they function in photosynthetic gas exchange. From measurements of O_2 and CO_2 levels in both fruits and seeds, it was concluded that photosynthesis could more than balance respiratory carbon loss. The presence of chloroplasts in a seed coat does not, however, automatically imply a photosynthetic CO_2 fixation. The inner seed coats of *Cucurbits* contain a green protochlorophyll pigment that is not converted to chlorophyll by light. Studies of mature seeds of *C. maxima* by Lott & Darley (1973) showed that these plastids have a quite unusual arrangement of thylakoids consisting of branched tubular membrane arrays sometimes attached to the inner plastid envelope. The pigment appears to be an unusual conversion product of normal chlorophyll and is not functional at seed germination. In *C. pepo*, Sundqvist & Ryberg (1979) found normal prolamellar bodies and starch grains in the inner coats of immature seeds. Their observations suggest that abnormalities resembling those that lead to chromoplast formation occur as part of advancing seed development. However, in chloroplasts of developing cotyledons of *Pisum sativum* and *Vicia faba* seeds, a high in vitro Hill activity has been reported, which is greater than that of leaves, as late as 35 days after anthesis (Banerji & Rauf, 1979).

A question arises whether all nutrition to the seed is supplied by the same donor organ. Using a 2 or 3 leaf plus one fruit system in the thorn-apple (*Datura*) with $K^{15}NO_3$ fed through the base of the cut stem and $^{14}CO_2$ supplied to the shoot, Lewis (1975) followed simultaneous supply to the fruit from the transpiration stream and the phloem. He concluded that the leaf in closest proximity to the fruit supplies nitrogen as well as photosynthate. From analysis of proteins and free amino acids in the seeds, leaf and petiole, it was clear that glutamine was the primary transported nitrogen assimilate passing to the seed. There, transamination reactions led to the ^{15}N enrichment of other amino acids and hence of protein. In somewhat comparable studies to determine the direction of flow of assimilates and sink-source inter-relations in soybean, Karpov & Kholupenko (1976) supplied ^{32}P and ^{14}C simultaneously. First $KH_2{}^{32}PO_4$ was applied in solution to the surface of chosen leaves, then $^{14}CO_2$ was introduced in the ambient gas phase of the same or different leaves. As most investigators in this field of physiology have found, the pod (in the axil) nearest to the labelled leaf is the recipient of most of the exported radioactivity. However, this situation changes during the period of 'mass plumping' when many of the pods along the axis are filling at the same time. By following the loss of ^{32}P and ^{14}C from leaves and determining the arrival of isotopes in the fruit, it was found that pods located at one tier of leaves may then receive assimilates from several other tiers of leaves but more from those above than those below the fruit. The greater the distance of a fruit from a donor leaf, the greater is the disparity between the carbon and phosphate intake. This non-equivalence in two transported assimilate classes is determined, they suggest, by the attracting power of the fruit and the heterogeneity of the leaves at the different stem tiers. Certainly, as a leaf ages or becomes senescent its physiological activity decreases and carbon fixation is reduced. Therefore, depending upon the location on the stem, the composition of photosynthetic products exported are also altered. The authors suggest that the compositional heterogeneity of the seed from a plant may in part derive from the diversity of products transported at each tier in the overall source-sink complex. What is not clear is whether fruit-sinks themselves exert any remote control over assimilate redistribution at positions where leaf traces fuse with the conducting elements of the stem, thereby ensuring admission to the recipient fruits of assimilates of a particular qualitative composition. They suggest that if this is so, then exchange between leaves and the fruits of different tiers should be particularly evident during the period of mass plumping. The importance of phloem in transport and differential movement of photosynthate and inorganic nutrients is emphasised from studies of 'split seed' disorder in *Lupinus* (Hocking et al., 1977). The discolouration, splitting and general deformity of the seed appears to result from low levels of manganese present in the phloem (bleeding sap) supplying fruit and seed. Though manganese is fully mobile in the xylem and is readily passed to centres of transpiration in the shoot, it exchanges only slowly with the phloem and is therefore poorly mobilised by the seeds. In this context the proposal that foliar senescence, or monocarpic senescence, results from a dilution of mineral nutrients in the leaves by mobilisation to the developing seeds seems unlikely. In experiments designed to test the concept, Sesay & Shibles (1980) made foliar applications of N, P and K in nutrient solution during the period of seed filling in soybean - a plant showing monocarpic senescence. Such treatment appeared not to delay the start or the course of degradative events of foliar abscission, though the rate at which they occurred was slower. Mobilisation of foliar reserves by the developing seed would not seem to be the deleterious event of senescence in the fruiting soybean.

An intriguing aspect of pod filling has been described in subterranean clovers where, after fertilisation, the inflorescences (the burrs) normally bury in the soil so that pod and seed development occur in the dark. If burrs are illuminated (Taylor, 1979) the weight of seeds per burr is markedly reduced; in some lines the reduction being linear with increases in the logarithm of light intensity. The mechanism by which pod size as well as seed size and seed numbers are reduced by illumination is unknown, nor is it clear whether the development of seed determines the growth of pods. Resolving the factors that determine transport of assimilate from donor organs to the seeds in *Trifolium subterranean* species is therefore a considerable challenge.

The loss of water from cereal and legume seeds at maturation is a remarkable process. The cells of most, if not all, of the maternal tissues are killed by the desiccation process. A distinction must be drawn between the monocotyledons, where the storage cells of the triploid endosperm rarely survive the drying experience though the embryo is fully desiccation-tolerant, and the dicotyledons, where most storage cells (cotyledons) are an integral part of the embryo and eventually reimbibe water to resume a fully functional metabolism. Clearly, major water loss must be via the seed coats(Rauf & Bannerji, 1978), but Rauf's (1978) analyses of total phosphorus and water contents in seeds of *Cicer, Pisum* and *Vicia* during desiccation indicate a back migration of water (including PO_4 in solution) to the rest of the plant. That water can redistribute within a single fruit in relation to the position of the seeds was shown by Sakiyama & Nakamura (1979) in stored cucumbers. When seeds were present at the distal end only, water was, in time, displaced to this region, whilst the seedless part below became pithy and degenerated. If the two halves were separated prior to storage or the fruits were nonseeded (parthenocarpic) no water movement or fruit degeneration occurred. It seems likely that salts in solution were also redistributed as water passed from one part of the fruit to another.

In the wild, water loss of grass and legume seeds prior to shedding generally reduces the moisture content to less than 20 %. In order to increase yields, some crop seeds (*Phaseolus vulgaris*) are harvested earlier in development at high moisture contents and then heat dried, often with deleterious effects on subsequent vigour at germination (Popa et al., 1978). In normal development, seed desiccation occurs only when the anatomical and biochemical differentiation has reached a precise and desiccation-resistant stage; in *Protea* species the period may be as long as 7 months (Van Staden, 1978a). Then, storage of seed in dry conditions and preferably at low temperatures is important to maintain the initial vigour (Van Staden, 1978b). At the other extreme, vivipary can occur, and is common in legumes where the embryo may grow continuously forming roots 10 cm long in the pod. This condition may well be related to a failure to desiccate at the right stage in development. In *Phaseolus vulgaris*, Pryke (1978) has associated the problem with high night temperatures at a particular stage in seed development. Equally important to embryo function after shedding are the conditions of hydration experienced during seed development. These can determine whether the shed seed shall be dormant or non-dormant at rehydration. In millet grown at high temperature, for example excised immature seeds are capable of germination as early as a week after flowering, and only when endosperm passes from the milk to the cellular stage is dormancy evident (Shimizu, 1978a).

The role of hormones in seed development

The role that hormones play in seed formation must be important not only in directing the complex differentiation pattern of embryo development but also by transducing changing environmental signals into biochemical switching systems, so regulating cell to cell communication and coordinated growth. There is a general problem however, in attempting to relate the levels of hormones in seeds or seed parts to some past, present or future state of seed development. Because cytokinins regulate cell division, particularly in tissue culture, they have been thought to play a role in determining the number of cell divisions and growth of cells in endosperms. Because green tissues treated with cytokinins act as loci for the accumulation of metabolites, it has been thought that one function of endogenous cytokinins in seeds might be to enhance the movement of assimilate to the seed. The possibilities are discussed by Holmes (1974) in relation to the function of awns in grain filling in barley, where awns are considered as potential sites of enhanced cytokinin levels. Certainly, in *Phaseolus coccineus* the major cytokinin changes from a biologically active but relatively non-polar form (zeatin), to polar but biologically poorly active conjugates (zeatin-glucoside or riboside) both in suspensor cells and in the embryo during the course of seed development - the active phases of cell growth coinciding with the highest levels of the more biologically active non-polar forms (Lorenzi et al., 1978). Because high growth rate and enhanced levels of zeatin are seen earliest in the suspensor, they conclude that the suspensor acts as a hormone source for the embryo. Previous work of this group identified relatively very high levels of gibberellin-like and auxin-like substances in the suspensor also. Doubtless, sugars, amino acid and other metabolite levels were high too.

Genetic mutants have proved of great value in helping to interprete hormone results. The use of a G2 pea mutant in which typical senescence is prevented in short days even though the plants flower, showed that levels of GA_{20} (the principal gibberellin of pea seeds) were higher in short days than in long days and rose when the seeds had passed the stage of maximum rate of cotyledon expansion. The levels of another gibberellin (GA_{29}) remained unmodified by photoperiod. From their data, obtained by GC-MS analyses and bioassay, Ingram & Browning (1979) argue that GA_{20} probably plays a role in embryogenesis and possibly also in the suppression of expansion growth of the developing cotyledons.

Photoperiod influences the levels of abscisic acid (ABA) during seed development as Makus & Shannon (1979) showed for snap bean (*Phaseolus vulgaris*). Developing seeds of plants in long days (16 h) had approximately 50 % higher ABA contents than in short days (10 h), independently of whether they were simultaneously exposed to high temperatures (35-46 °C). ABA and a number of ABA- derivatives in-

cluding phaseic acid, 4'-dihydrophaseic acid and 6-hydroxymethyl-abscisic acid have been identified by GC-MS from immature seeds of the cow pea, *Vigna unguiculata* (Adesomoju et al., 1980), but their role or activity in seed development is not proposed. In *Phaseolus*, however, using gas-liquid chromatographic quantification of ABA, Hsu (1979a) found free and bound ABA to be low in pods, but free ABA to be high in seed coats from anthesis and throughout the period of rapid seed growth. In embryos, free ABA was also high during rapid expansion growth and during maximum storage protein deposition. Clearly, ABA cannot be functioning as a growth inhibitor at either of these stages. As explanation Hsu suggests that the mRNA's for storage protein synthesis are transcribed early and that ABA selectively inhibits most protein synthesis whilst permitting translation of messages for storage protein synthesis to proceed.
Observing the effects of different hormones on embryos isolated from wheat at different stages of seed development Radley (1979) also concluded that at stages when germination of grains could be achieved, large amounts of free endogenous ABA were present. Since added ABA inhibited germination, it seemed likely that the endogenous ABA might be compartmentalised and hence unavailable to the embryo. She notes too that IAA levels declined during this period, but added IAA (as high as 40 μg/ml) induced no physiological response. It is evident from such studies that we still have much to learn before we gain an understanding of the role of hormones during seed formation and particularly we need to know why cells of embryo and endosperm respond to hormones in different ways from the rest of the plant. The high rate of growth and storage deposition when ABA levels are high is an example, for ABA causes growth inhibitions in other plant parts. The claim that storage deposition is stimulated in sunflower, poppy and cotton by late application of another growth inhibitor, maleic hydrazide, may therefore be relevant (Zhdanova & Rusova, 1979) though the paper (in translation) is difficult to interpret.

Biochemical changes during seed development

The concept that clones of cells could be important in determining the differentiation of tissue types was set out by McClintock (1978). The relevance of relating cell lineages to clonal distribution of organelle-inherited cytoplasmic properties is evident; for example, some unstable traits such as chloroplast mutator or non-chromosomal stripe appear consistent with cell lineage derivation (Coe & Neuffer, 1978). Histochemical distinctions of specific tissue areas of embryos are few but have been attempted in developing seeds of *Panicum* (Rudramuniyappa & Panchaksharappa, 1978). From the quadrant embryo stage a histochemical demarcation was apparent between the two poles: cells destined to become the embryo proper showing intense RNA and protein synthesis with high levels of lipids and reducing sugars, whilst the basal area (which gives rise to the suspensor tissue) had consistently low levels. An example of a more precise biochemical tissue marker comes from legume seeds. Certain specific protein inhibitors of proteases are restricted to the seed. Kapoor & Gupta (1978) found that 83 % of the trypsin inhibitor of soybean could be accounted for in the cotyledons, the activity apparently initiated early in the milk stage. No activity was detectable in the embryo axis.

The localised production of specific proteins, lipids, or polysaccharides by cells of either cotyledons or endosperm, characterise their biochemical identity and clonal differentiation during development. At all stages of development from the time of pollination to mature kernels, endosperms of *Zea mays* produce a specific hydrophobic protein, zein. Fractionation of extracts of tissue by electrophoresis in polyacrylamide SDS or urea gels to determine electrophoretic mobilities has shown (Rewa & Brückner, 1979) that the zein molecule is present as eight subunits associated as twelve aggregates throughout grain development. Proteins (globulins) synthesised as storage components in legume seeds show considerable species specificity. In the cotyledons of the kidney bean, *Phaseolus vulgaris*, the major storage protein is a globulin somewhat analogous to the vicilin and legumin of *Pisum sativum* and *Vicia faba*. The bean globulin has been extracted and purified by Sephadex chromatography and polyacrylamide gel fractionation. Molecular weight determinations of the peptide bands in comparison with standards has convincingly demonstrated the different identity of the *Phaseolus* globulin (called Euphaseolin) from vicilin and legumin (Murray & Crump, 1979). Immunoprecipitation assays by double-diffusion against concanavalin A confirmed that the preparation of Euphaseolin was a concanavalin A-interacting glycoprotein.

It is evident that cells of tissues that synthesise storage protein must contain many copies of the mRNA that codes for them. After desiccation the dry seeds contain mRNA molecules that will code for protein in cell free protein synthesising systems. The question has been debated, do the mRNA-containing ribonucleoprotein particles (or informosomes) retain function in the dry seed and are they translated again at germination? Peumans et al. (1979) followed the synthesis of informosomes in developing rye embryos until maturity and showed that though total mRNA content per embryo increased with growth of the embryo, the proteins coded for remained qualitatively identical as judged from products synthesised in wheat embryo cell free systems. Since embryos excised early in their development still germinate effectively, most of the mRNA present as informosomes studied in mature seed must be unessential for germination. Because storage proteins are not synthesised by cells other than those of storage tissues of endosperm or cotyledon, it seems that any residual mRNA coding for those proteins must either be degraded at germination or have only a transitory translational life. If the ribonuclease activity usually associated with extracted ribosomes in vitro (Siwecka et al., 1979) were active in vivo at germination, then loss of stored mRNAs at germination might be explained.

In many leguminous species seeds of the first generation hybrid cross between two varieties have characteristics resembling those of the maternal parent, particularly in respect to their seed storage proteins, the globulins, where selective expression of alleles derived from the maternal genome can occur (Davies, 1973). Globulins of the F1 generation in pea cotyledons showed electrophoretic patterns identical with the maternal parent, with very little, if any, of the paternal parent detectable. Several hypotheses could be proposed for this preferential accumulation of maternal-type storage protein. Paternal globulin products could be preferentially degraded or maternal globulins preferentially transcribed by long-lived mRNA produced by the maternal parent. Alternatively, there could be repression of paternal genes or activation of maternal genes during seed development. Selective repression or activation of specific loci in the hybrid genome is most favoured since globulins are produced only at a precise and somewhat late stage in embryo development. It seems reasonable of Davies to ascribe the maternal similarity to a maternal activation rather than a paternal repression; the maternal parent producing an inducer which is recognised only by the repressor locus of the maternal chromosomes. In the hybrid embryo the maternal structural gene would then produce maternal-type globulin. If the 'inducer' were to be preferentially recognised or have a higher affinity for the maternal suppressor, higher production of maternal globulin might arise in all crosses. The genetic control of seed protein characters in storage cells, as exerted by the maternal parent, has obvious and important implications in seed protein production.

Briarty (1978) reviews much of the history of the unravelling of mechanisms of protein deposition in membrane bound protein bodies in the storage cells of seeds of major crop legumes and cereals. He discusses an important aspect of the deposition process, namely how protein synthesised on the rough endoplasmic reticulum is transported to the inside of the protein body. Norton et al. (1978) review the anatomy of the development of legume seeds, calling attention to the transitory storage function of seed integuments and endosperm as distinct from the major storage function of the specialised cells of the embryo (the cotyledons). He raises another important question, namely, what biochemical events determine the critical stage in embryo development when storage protein is 'switched on'? Bollini & Chrispeels (1979) have supplied new information to these questions. In studies of reserve protein synthesis in *Phaseolus vulgaris* cotyledons, they investigated a) sites of synthesis of vicilin (but see comments of Murray & Crump, 1979) and phytohaemagglutinin and b) the mechanism of accumulation of the protein in the protein bodies. From sucrose density gradient experiments, membrane bound polysomes attached to the endoplasmic reticulum were confirmed as sites of in vitro vicilin synthesis. Some of the peptides so formed were identified by their mobility in SDS polyacrylamide gels and also by immunological precipitin reactions when challenged with vicilin antiserum. Since there appears to be no continuity between endoplasmic reticulum and the membrane of the protein body, a mechanism for protein transfer to the inside would appear to involve specific transmembrane processes. The addition of small sugar moieties to form glycoproteins may facilitate this transfer. In relation to this, Davies & Delmer (1979) incubated particulate preparations from developing cotyledons of *Phaseolus vulgaris* which, when supplied with radioactive uridine 5'-diphospho-N-acetyl-D-glucosamine, led to glycosylation of associated protein and the production of vicilin and phytohaemagglutinin as labelled products; the vicilin component reacting with antiserum to vicilin extracted from mature seeds. In a thorough study of the isolated sub-units of storage globulins in *Lupinus angustifolius* (Eaton-Mordas & Moore, 1979), mannose, galactose and glucosamine were shown to be present. Heterogeneity of the polypeptide chains was suspected both in sugar and oligosaccharide content, which may reflect a glycosylation or proteolytic modification during embryo development. The role of glycosylation at seed protein deposition in protein bodies and the autophagous nature of these bodies during mobilisation of reserves at germination is still incompletely resolved (Dhar & Vijayaraghavan, 1979).

Certain seeds accumulate high levels of polysaccharide of particular composition in the endosperm. In *Gleditsia triacanthos*, for example, mature cells contain predominantly three galactomannans of varying mannose-galactose ratios, but with small amounts of protein attached (Mazzini & Cerezo, 1979). This suggests the presence of glycoproteins in which the sugar moieties greatly outnumber the amino acid components. In *Cassia corymbosa* seed, which contains high amounts of galactomannans, enzymes converting sugars of the D-gluco-configuration to manno-configurations have been isolated and purified. Small & Matheson (1979) report the complete separation and kinetic properties of a phosphomannomutase and a phosphoglucomutase from developing seeds and propose that these enzymes operate by converting transported sugars from photosynthate into the mannan-component of storage polysaccharide in the seed.

Although protein and carbohydrate comprise the major storage reserves of most monocotyledon seeds, many dicotyledons, particularly legumes, accumulate lipids and free fatty acids. Lipid synthesis is continuous in all actively metabolizing cells, but only specialized cells are directed to lipid deposition. In comprehensive analyses with quantification of lipid in the embryo, endosperm and pericarp tissues of developing maize kernels, Tan & Morrison (1979) drew attention to the storage role of accumulated lipid during the immature stages. During early embryogenesis, triglycerides accumulate in sphaerosomes of embryo cells, but at maturity the embryo contains most of the kernel's phospholipid with little free fatty acid, mono- or di-glyceride or other lipid degradation products. This would accord with the maintenance of membrane integrity and the respiratory utilisation of triglycerides during embryo maturation . It would also account for the general absence of lipid degradation products. The lipid status of aleurone is somewhat similar to that of the embryo. In the endosperm, however, most lipids are synthesised and accumulate during the first 23 days of development after pollination. When cell division ceases, levels decrease and by 36 days after pollination phospholipid degradation is extensive. This poses problems since phospholipids are essential to membrane integrity and therefore to the continued synthesis of storage protein and to the maintained compartmentation of protein bodies and amyloplasts.

Lipid biosynthesis has been studied in considerable detail in safflower, *Carthamus tinctorius* by Ichihara & Noda (1980). They followed the changes in the fatty acid components of triacylglycerols, phospholipids and glycolipids during seed development and investigated ^{14}C-acetate incorporation into total lipid products in vitro, by seeds excised from plants at different developmental stages.

During the period of rapid growth and active lipid synthesis, the labelled fatty acids were predominantly unsaturated. This is not surprising since the dienoic linoleic acid comprises about 70 % of the total fatty acids in safflower seeds. In mature seeds, however, saturated or monoenoic fatty acids were predominantly labelled. Interestingly, raising the temperature of the incubation medium led to a higher proportion of saturated fatty acids amongst the labelled products. The fatty acid composition of safflower oil in the field plants is similarly temperature dependent, (Harris et al., 1978) and high night temperatures during growth lead to reductions in total lipid yields. The availability of oxygen in vitro may be responsible for the incorporation results obtained by Ichihara & Noda (1980) with excised seeds, but they also indicate possible similarities with lipid biosynthesis in vivo.
Perhaps one of the most unusual lipids synthesised from seeds is the liquid 'wax' of the jojoba, *Simmondsia chinensis* (Rost et al., 1977). The cotyledon cells are filled with liquid lipid droplets which comprise 48 % of the weight of the seed. At germination it is presumed that the lipid is degraded and the products transported from the cotyledons and utilised for growth by the embryo axis. At no time does the axis itself contain the extractable wax.
Protein bodies do not, of course, contain only protein lipid and polysaccharides. Most of the mineral reserves of P, K, Ca and Mg and minor metallic elements are present as salts of hexaphosphoric acid (phytin) in globoid crystalline inclusions. The mineral deposition in cells of storage cotyledons of *Astragalus bisulcatus* and *Oxytropos lambertii* is as distinct from the rest of the plant as the proteins they synthesise (Lott & Vollmer, 1979). For example, these plants accumulate selenium ions in their vegetative tissues, but energy-dispersive X-ray (EDX) analysis shows the selenium is completely excluded from the protein bodies, emphasising the special biochemical character of the seed storage cells. An EDX study of the composition of phytin globoids in aleurone cells of rice grains (Ogawa et al., 1979) showed that most major and minor mineral components were deposited but with a delayed or varied accumulation of K. Interestingly, P, Mg and K are not found in protein bodies of the endosperm.

The seed is the start
the seed is the end
and what takes place between the two
is biochemical mystery.

References

Abdel-Rahman, A-H.Y. & S.A.M. Youssef, 1977. Protein level in developing seeds of Vicia faba L. and their quality in relation to pod position on the stems. Journal of Food Technology 12: 485-491.
Adesomoju, A.A., J.I. Okogun, D.E.U. Ekong & P. Gaskin, 1980. GC-MS identification of abscisic acid and abscisic acid metabolites in seed of Vigna unguiculata. Phytochemistry 19: 223-225.
Atkins, C. & A. Flinn, 1978. CO_2 fixation in the carbon economy of developing seeds of Lupinus albus L. Plant Physiology 62: 486-490.

Balatková, V., J. Tupý & E. Hrabetová, 1977. Seed formation in Narcissus pseudonarcissus L. after placental pollination in vitro. Plant Science Letters 8: 17-21.

Banerji, D. & A. Rauf, 1979. Comparative growth and biochemical studies on seed development. 3. Chlorophyll development and Hill activity in the developing seeds of Pisum sativum and Vicia faba. The Plant Biochemical Journal 6: 31-35.

Bennett, M.D., 1974. Meiotic, gametophytic, and early endosperm development in triticale. In: R. MacIntyre and M. Campbell (Eds), Triticale: proc. int. symp., El Baton, Mexico, Int. Develop. Res. Centre Monogr. IDRC-024e, Ottawa, Ontario. p. 137-148.

Bhatnagar, S.P. & S. Chandra, 1976. Reproductive biology of Abelmoschus esculentus. VIII. Seed disfunction. Botanische Jahrbücher 97: 425-439.

Bollini, R. & M.J. Chrispeels, 1979. The rough endoplasmic reticulum is the site of reserve-protein synthesis in developing Phaseolus vulgaris cotyledons. Planta 146: 487-501.

Bopp, M., 1979. VII. Developmental physiology. Progress in Botany 41: 135-152.

Bradley, M.V. & J.C. Crane, 1975. Abnormalities in seed development in Pistacia vera L. Journal of the American Society for Horticultural Science 100: 461-464.

Brar, G. & W. Thies, 1977. Contribution of leaves, stem, siliques and seeds to dry matter accumulation in ripening seeds of rapeseed, Brassica napus L. Zeitschrift für Pflanzenphysiologie 82: 1-13.

Breidert, D. & W.J. Schön, 1979a. Die Bildung und Speicherung von Kohlenhydraten und Proteinen in reifenden Gerstenkaryopsen. I: Methodenübersicht, Kohlenhydrate und Stickstoff-Fraktionen. Angewandte Botanik 53: 65-81.

Breidert, D. & W.J. Schön, 1979b. Die Bildung und Speicherung von Kohlenhydraten und Proteinen in reifenden Gerstenkaryopsen. II. Angewandte Botanik 53: 145-159.

Briarty, L.G., 1978. The mechanisms of protein body deposition in legumes and cereals. In: G. Norton (Ed.), Plant Proteins. Butterworths, London. p. 81-106.

Chatterjee, S.R., T.C. Pokhriyal & Y.P. Abrol, 1978. Changes in the content of protein and amino acids in developing pea seeds. The Plant Biochemical Journal 5: 69-76.

Coe, E. & M. Neuffer, 1978. Embryo cells and their destinies in the coru plant. 36th Symp. of Soc. for Dev. Bot. The clonal basis for development. p. 113-129 (summary only).

Crane, J.C., 1975. The role of seed abortion and parthenocarpy in the production of blank pistachio nuts as affected by rootstock. Journal of the American Society for Horticultural Science 100: 267-270.

Davies, D.R., 1973. Differential activation of maternal and paternal loci in seed development. Nature 245: 30-32.

Davies, H.M. & D.P. Delmer, 1979. Seed reserve-protein glycosylation in an in vitro preparation from developing cotyledons of Phaseolus vulgaris. Planta 146: 513-520.

Dhar, U. & M. Vijayaraghavan, 1979. Outogeny, structure and breakdown of protein bodies in Linum usitatissimum L. Annals of Botany

43: 107-111.
Diepenbrock, W. & G. Geisler, 1979. Compositional changes in developing pods of oilseed rape (Brassica napus L.) as affected by pod position on the plant. Canadian Journal of Plant Science 59: 819-830.
Eaton-Mordas, C.A. & K.G. Moore, 1979. Subunit Glycosylation of Lupinus angustifolius seed globulins. Phytochemistry 18: 1775-1779.
Gengenbach, B.G., 1977. Development of maize caryopses resulting from in vitro pollination. Planta 134: 91-93.
Grear, J.W. & N.G. Dengler, 1976. The seed appendage of Eriosema (Fabaceae). Brittonia 28: 281-288.
Harris, H.C., J.R. McWilliam & W.K. Mason, 1978. Influence of temperature on oil content and composition of sunflower seed. Australian Journal of Agricultural Research 29: 1203-1212.
Hocking, P.J. & J.S. Pate, S.C. Wee & A.J. McComb, 1977. Manganese nutrition of Lupinus spp. especially in relation to developing seeds. Annals of Botany 41: 677-688.
Holmes, D.P., 1974. Physiology of grain filling in barley. Nature 247: 297-298.
Hsu, F.C., 1979a. Abscisic acid accumulation in developing seeds of Phaseolus vulgaris L. Plant Physiology 63: 552-556.
Hsu, F.C., 1979b. A development analysis of seed size in common bean. Crop Science 19: 226-230.
Ichihara, K-I & M. Noda, 1980. Fatty acid composition and lipid synthesis in developing safflower seed. Phytochemistry 19: 49-54.
Ingram, T.J. & G. Browning, 1979. Influence of photoperiod on seed development in the genetic line of peas G2 and its relation to changes in endogenous gibberellins measured by combined gas chromatography - mass spectrometry. Planta 146: 423-432.
Jassem, M., 1973. Endosperm development in diploid, triploid and tetraploid seed of sugarbeet (Beta vulgaris L.). Genetica Polonica 14: 295-303.
Jernstedt, J.A. & C. Clark, 1979. Stomata on the fruits and seeds of Eschscholzia (Papaveraceae). American Journal of Botany 66: 586-590.
Kapoor, A.C. & Y.P. Gupta, 1978. Trypsin inhibitor activity in soybean seed as influenced by stage of its development and different treatments, and the distribution in its anatomical parts. Indian Journal of Nutrition and Dietetics 15: 429-433.
Karpov, E.A. & I.P. Kholupenko, 1976. Comparative study of transport of radioactive isotopes of phosphorus and carbon from soybean leaves during the period of seed plumping. Soviet Plant Physiology 23: 852-858.
Konno, S., 1979. Changes in chemical composition of soybean seeds during ripening. Japan Agricultural Research Quarterly 13: 186-194.
Korohoda, J. & K. Strzalka, 1979. High efficiency genetic transformation in maize induced by exogenous DNA. Zeitschrift für Pflanzenphysiologie 94: 95-99.
Kumar, U., 1977. Morphogenetic regulation of seed germination in Orobanche aegyptiaca Pers. Canadian Journal of Botany 55: 2613-2621.

Lewis, O.A.M., 1975. An ^{15}N-^{14}C study of the role of the leaf in the nitrogen nutrition of the seed of Datura Stramonium L. Journal of Experimental Botany 26: 361-366.

Lorenzi, R., A. Bennici, P.G. Cionini, A. Alpi & F.D'Amato, 1978. Embryo-suspensor relations in Phaseolus coccineus: cytokinius during seed development. Planta 143; 59-62.

Lott, J.N.A. & J.J. Darley, 1973. Unusual thylakoid arrangements in plastids from Cucurbita maxima seed coat cells. Cytobiologie 8: 55-60.

Lott, J.N.A. & C.M. Vollmer, 1979. An energy dispersive x-ray analysis study of protein bodies in the seeds of Astragalus bisulcatus. and Oxytropis lambertii. Canadian Journal of Botany 57: 987-992.

Lucas, E.O. & G.M. Milbourn, 1979. The role of pre-anthesis assimilate in the seed development of Phaseolus vulgaris. Journal of Nuclear Agriculture and Biology 8: 81-84.

McClintock, B., 1978. Development of the maize endosperm as revealed by clones. 36th Symp. of Soc. for Dev. Bot. The clonal basis of development. p. 217-237.

Makus, D.J. & C. Shannon, 1979. Temperature and photoperiod effects on abscisic acid content in 'Earliwax' Snapbean seed. Horticultural Science 14: 732-733.

Martinez-Carrasco, R. & G.N. Thorne, 1979. Physiological factors limiting grain size in wheat. Journal of Experimental Botany 30: 669-679.

Mazzini, M.N. & A.S. Cerezo, 1979. The carbohydrate and protein composition of the endosperm, embryo and testa of the seed of Gleditsia triacanthos. Journal of the Science of Food and Agriculture 30: 881-891.

Murray, D.R., 1979. Nutritive role of the seedcoats during embryo development in Pisum sativum L. Plant Physiology 64: 763-769.

Murray, D.R. & J.A. Crump, 1979. Euphaseolin, the predominant reserve globulin of Phaseolus vulgaris cotyledons. Zeitschrift für Pflanzenphysiologie 94: 339-350.

Norton, G., J.F. Harris & A. Tomlinson, 1978. Development and deposition of protein in oilseeds. In: G. Norton (Ed.), Plant Proteins. Butterworths, London. p. 59-79.

Ogawa, M., K. Tanaka & Z. Kasai, 1979. Energy-dispersive X-ray analysis of phytin globoids in aleurone particles of developing rice grains. Soil Science and Plant Nutrition 25: 437-448.

Openshaw, S.J., H.H. Hadley & C.E. Brokoski, 1979. Effects of pod removal upon seeds of nodulating and nonnodulating soybean lines. Crop Science 19: 289-290.

Peumans, W.J., L.I. Caers & A.R. Carlier, 1979. Some aspects of the synthesis of long-lived messenger ribonucleoproteins in developing rye embryos. Planta 144: 485-490.

Popa, G., L.O. Copeland & M.W. Adams, 1978. Physiological development of seed of two dry edible bean varieties. Annual Report of the Bean Improvement Cooperative 21: 41-44.

Pryke, P.I., 1978. Dormancy failure in bean seeds. Annual Report of the Bean Improvement Cooperative 21: 24-25.

Radley, M., 1978. Factors affecting grain enlargement in wheat. Journal of Experimental Botany 29: 919-934.

Radley, M., 1979. The role of gibberellin, abscisic acid, and auxin

in the regulation of developing wheat grains. Journal of Experimental Botany 30: 381-389.
Rauf, S., 1978. Comparative growth and biochemical studies on seed development. 1. Biochemical changes in developing seeds and pods of certain legumes. The Plant Biochemical Journal 5: 157-166.
Rauf, A. & D. Banerji, 1978. Comparative growth and biochemical studies on seed development. 2. Biochemical changes in developing seed parts of certain legumes. The Plant Biochemical Journal 5: 167-173.
Rewa, W.A. & C. Brückner, 1979. Das Molekulargewicht des Zeins und sein Verhalten bei der Äthanol-Gradientextraktion in verschiedenen Reifestadien der Samen. Biochemie und Physiologie der Pflanzen 174: 462-467.
Rost, T.L., A.D. Simper, P. Schell & S. Allen, 1977. Anatomy of jojoba (Simmondsia chinensis) seed and the utilisation of liquid wax during germination. Economic Botany 31: 140-147.
Rudramuniyappa, C.K. & M.G. Panchaksharappa, 1978. Histochemical study of seed development in Panicum. Beiträge zur Biologie der Pflanzen 54: 165-173.
Sakiyama, R. & S. Nakamura, 1979. Rate of weight loss in proximal and distal halves of picked cucumber fruits as affected by developing seeds. Scientia Horticulturae 11: 349-355.
Salminen, S.O. & R.D. Hill, 1978. Rate and duration of dry matter accumulation in relation to kernal shrivelling in Triticale. Canadian Journal of Plant Science 58: 13-19.
Sesay, A. & R. Shibles, 1980. Mineral depletion and leaf senescence in soya bean as influenced by foliar nutrient application during seed filling. Annals of Botany 45: 47-55.
Shimizu, N., 1978a Studies on dormancy and germination of seed in finger millet (Eleucine coracana). III. Seed development and ripening and changes in dormancy and germination. Bulletin of the National Grassland Research Institute (Sochi Shikenjo Kenkyu Hokoku) 13: 24-40 (Eng. summary).
Shimizu, N., 1978b. Development and germination behaviour of seed produced by culture of cut ear pasture plants. I. The possibility of seed development by culture of cut ear in temperate grasses. Journal of the Japanese Society of Grassland Science 23: 295-300 (Eng. summary).
Siwecka, M.A., M. Rytel & J.W. Szarkowski, 1979. The presence of deoxyribonucleolytic activity in cytoplasmic ribosomes of rye (Secale cereale L.) germs. Acta Biochimica Polonica 26: 97-101.
Small, D.M. & N.K. Matheson, 1979. Phosphomannomutase and phosphoglucomutase in developing Cassia corymbosa seeds. Phytochemistry 18: 1147-1150.
Sokolov, I.D. & Y.P. Kramarenko, 1974. Abnormalities of cell formation in the endosperm of Iris pseudacorus L. Cytology and Genetics 8: 59-62.
Staden, J. van, 1978a. Seed viability in Protea neriifolia. I. The effects of time of harvesting on seed viability. Agroplantae 10: 65-67.
Staden, J. van, 1978b. Seed viability in Protea neriifolia. II. The effects of different storage conditions on seed longevity. Agroplantae 10: 69-72.

Staritsky, G., 1973. Seed morphology and early development of the ground-nut (Arachis hypogaea L.). Acta Botanica Neerlandica 22: 373-389.

Stebbins, G.L., 1976. Seeds, seedlings and the origin of angiosperms. In: C.B. Beck (Ed.), Origin and Early Evolution of Angiosperms. p. 300-311.

Sundqvist, C. & H. Ryberg, 1979. Structure of protochlorophyll - containing plastids in the inner seed coat of pumkin seeds (Cucurbita pepo). Physiologia Plantarum 47: 124-128.

Tan, S.L. & W.R. Morrison, 1979. Lipids in the germ, endosperm and pericarp of the developing maize kernel. Journal of the American Oil Chemists' Society 56: 759-764.

Taylor, G.B., 1979. The inhibitory effect of light on seed and burr development in several species of Trifolium. Australian Journal of Agricultural Research 30: 895-907.

Turała-Szybowska, K., 1974. Karyological anatomy of the endosperm of Bryonia dioica and Cucurbita pepo. Acta Biologica Gracoviensia, Ser: Botanica 17: 75-90.

Williams, E. & D.W.R. White, 1976. Early seed development after crossing of Trifolium ambiguum and T. repens. New Zealand Journal of Botany 14: 307-314.

Yadav, N.S., K. Postle, R.K. Saiki, M.F. Thomashow & M.D. Chilton, 1980. T-DNA of a crown gall teratoma is covalently joined to host plant DNA. Nature 287: 458-461.

Youssef, S.A.M. & A-H.Y. Abdel-Rahman, 1977. Protein level in developing seeds of Vicia faba L. and quality in relation to pod's position on the stems. L'agricoltura Italiana 106: 169-178.

Zatyko, J.M., I. Simon, C. Szabó & P. Gracza, 1979. Histological and chemical analysis of red currant seeds developing in vitro. Acta Agronomica Academiae Scientiarum Hungaricae 28: 313-319.

Zhdanova, L.P. & M.I. Rusova, 1979. Effect of maleic hydrazide on accumulation of storage substances in seeds. Soviet Plant Physiology 26: 341-347.

Changes in seed composition during germination

J.C. Slaughter

Carbohydrates

Pal et al. (1976) found that germination of bajra at 25 °C and barley at 15 °C gave a better yield of malt than obtained at higher temperatures. The malts produced at the lower temperatures had much higher β-amylase activities. The production of amylase by half seeds of barley under the influence of gibberellic acid (GA_3) was reduced by tannins apparently acting in competition with the hormone (Jacobson & Corcoran, 1977). Also using a half-corn system, Rodaway & Kende (1978) found that, contrary to previously published work, 5-fluorouracil had no effect on the properties of α-amylase produced during barley germination. Morrall & Briggs (1978) studied the germination of decorticated barley seeds over a 6-d period and concluded that endosperm cell walls provided 20 % of the degradable carbohydrate reserve.

In germinating wheat, Goswami et al. (1977) showed that hydrolysis of starch by β-amylase began at once and α-amylase only became the dominant enzyme after 2 d. Daussant & Hill (1979) showed that the α-amylase system of germinating *Triticale* seeds was similar to that of wheat.

Okamoto & Akazawa (1978) purified an α- and a β-amylase from the endosperm of rice seeds which had been germinated for 7 d at 30 °C. Igaue et al. (1978) also reported purification of an α-amylase from germinated rice. In the period before starch breakdown, Okamoto & Akazawa (1979) showed that the amylase activity of rice was located almost entirely in the epithelium septum between the scutellum and endosperm. Only in the later stages of germination did enzyme develop in the aleurone cells. Using [^{35}S] methionine incorporation, Okamoto & Akazawa (1980) obtained evidence that β-amylase is initially synthesised in the scutellum of rice, but that activation of the inert form associated with the starch granules becomes the dominant source of enzyme later in germination. In winter rye germinated for 24 h, Wilp & Buschbeck (1978) showed that the four main α-amylase isoenzymes were present in embryo, scutellum and in the remainder of the caryopsis. In the embryo or scutellum β-amylase was not present.

The α-amylase activity of *Phaseolus vidissimus* Ten. increased markedly over a 5-d germination period and was located almost entirely in the cotyledons (Yamasaki & Suzuki, 1979). The enzyme has been highly purified. The main amylase in germinating pea seed is β-amylase. Davis (1977) found α-amylase activity initially only in the hypocotyl but, as germination proceeded, the epicotyl became the main location of the enzyme. During germination of *Phaseolus*

mungo seeds, α-amylase increased in the cotyledons in the areas furthest away from the vascular bundles (Minamikawa & Koshiba, 1979). The starch content of these areas also decreased selectively.
Zimniak-Przybylska & Przybylska (1976) noted that the storage proteins of a wide range of *Pisum* varieties could interfere with the starch-iodine reaction used in producing amylase zymograns. This can result in erroneous results being obtained on electrophoresis of amylases. Tárrago & Nicolás (1976) found that the degradation of starch in germinating lentil cotyledons correlated better with α-amylase activity than with β-amylase or phosphorylase. The same conclusion was reached by Umezurike & Numfor (1979) for the tropical pulse, *Voandzeia subterranea* L. and for azuki seeds by Hizukuri & Nakamura (1978). The kinetic properties of two starch phosphorylases from germinated pea seeds have been described by Matheson & Richardson (1978).

Ballance et al. (1976) found that the specificity of β-glucanase activity in barley seeds depended on the tissue examined. Endo-β-1,3-glucanase occurred in the embryo and scutellum of unsoaked seeds, but increased markedly in the aleurone and endosperm during germination. Bacic et al. (1977) found that β-1,3-glucanase activity barely increased at all during germination of *Stylosanthes humilis* whereas amylase and invertase showed distinct rises.

The activity of α-galactosidase and pentosanase increased greatly during the first 3 d of germination of groundnuts and the concentrations of raffinose, stachyose and verbascose decreased (Wankhede et al. (1977). Glucose, fructose, sucrose and starch all increased markedly. The activity of both α- and β-galactosidase increased during germination of *Pisum sativum* and *P. elatius* (Marbach et al., 1978). Seiler (1977) showed that the galactomannan reserves of the endosperm of carob seeds begin to be hydrolysed once the radicle penetrates the seed coat. The synthesis of the hydrolytic enzymes is not under control of the embryo. Reid et al. (1977) demonstrated that an endo-β-mannanase is involved in the breakdown of endosperm reserves in germinating fenugreek seeds while according to Stoech & Grubert (1978) galactose is preferentially removed from the galactomannan of the mucilaginous endosperm of *Cercis siliquastrum* during germination. Bouquelet & Spik (1978) have isolated 4 forms of N-acetyl-β-D-hexosaminidase from different parts of germinating fenugreek seeds. The thick, mannan-rich endosperm cell walls of lettuce seeds are broken down during germination by an endo-β-mannase synthesised in the endosperm (Halmer et al., 1978). Abscisic acid inhibits production of this enzyme by isolated endosperm (Halmer & Bewley, 1979). Leung et al. (1979) have suggested that completion of mannan hydrolysis in germinating lettuce seed occurs in the cotyledon as this is the site of β-mannosidase activity. During the germination of lupin seeds (Matheson & Saine, 1977), bambarra groundnut seeds (Amuti & Pollard, 1977) chick peas and mung beans (Aman, 1979) there is a rapid decrease in the concentration of raffinose oligosaccharides associated with a temporary rise in sucrose content. The embryos of germinating fenugreek seeds take up the galactose and mannose liberated from the endosperm entirely by passive diffusion (Uebelmann, 1978).

Palmitoyl-CoA: *sn*-glycerol 3-phosphate acyltransferase, an enzyme necessary for the formation of membrane precursors, rose to a maximum activity at the end of imbibition of castor bean seeds (Vick & Beevers, 1977). The enzyme was localized on the endoplasmic reticulum of the endosperm. Donaldson & Beevers (1978) purified endoplasmic reticulum of germinating castor beans and biochemically characterized the fractions. Yamada et al. (1978) demonstrated the presence of a factor which could transfer phospholipid from microsomes to mitochondria in the same tissue. Tanaka & Yamada (1979) isolated a phospholopid exchange protein from castor beans which functioned with microsomes and mitochondria from a range of species.

Allfrey & Northcote (1977) studied the breakdown of reserve lipid during germination of groundnuts and the pattern of accumulation of the low molecular weight products. Huang & Moreau (1978) showed that the pattern of lipase development during germination of castor beans is not shared by other oil seeds. Peanuts contain an alkaline lipase which is located mainly in the glyoxysomes and is absent from the lipid bodies. Yoshida & Kajimoto (1977a, b) showed that polar lipids came to predominate over non-polar lipids in the axis of soybeans during germination. The lipase activity of sunflower seeds increased steadily during germination (Bhatia et al., 1978) and there was a rapid decline in the total lipid content of the endosperm. A histochemical examination of germinating seeds of *Quercus alba* carried out by Vozzo (1978) showed a transfer of lipid from the cotyledons to the embryonic axis.

During imbibition, lima beans incorporated a higher proportion of ^{14}C-glycerol into phosphatidyl-ethanolamine and phosphatidylglycerol than did peas or broad beans. At the seedling stage incorporation was the same in all four species (Dogras et al., 1977). Yoshida & Kajimoto (1977a) observed that in the germination of soybean seeds non-polar lipids were metabolized faster than polar lipids. Yoshida & Kajimoto (1978) described the fatty acid distribution of glycolipids and phospholipids in germinated soybean cotyledons. Wharfe & Harwood (1979) purified an ethanolamine kinase from germinating soybean separate from choline kinase.

The major sterols of longleaf pine were campesterol and β-sitosterol and in germinating seeds the total sterol concentration increased over a 20-d period (Van Vu & Biggs, 1978) whereas with fenugreek seeds a maximum occurred at 6 d (Lockwood & Brian, 1976). Duperon et al. (1979) also noted that the total sterol concentration in lettuce seeds reached a maximum at about 10 d. Van Vu & Biggs (1979) reported investigations into the effect of tris-(2-diethylaminoethyl)-phosphate trihydrochloride, abscisic acid and cychloheximide on sterol biosynhesis during the germination of longleaf pine. Doireau (1977) showed that β-amyrin synthesis was higher than that of free sterols in germinating pea cotyledons and that β-amyrin increased throughout the developing plant during germination. Maudinas & Lematre (1979) found that the major yellow pigment synthesised in the root caps of maize during germinating was violaxanthin.

Polyphenols and oxidases

The metabolism of sinapolycholine was studied in four *Cruciferae* species (Strack et al., 1978) and in three cases the levels declined during germination, but a complex pattern was found with *Raphanus sativus*. Krause (1978) showed that the endosperm was necessary for the formation of flavanoids in the emerging seedlings of *Fagopyrum esculentum*.
Minamikawa (1977) purified quinate: nicotinamide adenine dinucleotide (NAD) oxidoreductase from 2-d-old etiolated seedlings of *Phaseolus mungo*. The large increase in peroxidase activity in the cotyledons of germinating *Phaseolus aureus* seedlings observed by Dendsay & Sacher (1978) was strongly inhibited by abscisic acid. This inhibition was overcome by addition of auxin. Abscisic acid also markedly inhibited the development of indoleacetic acid (IAA) oxidase and peroxidase during the germination of rice (Sengupta et al., 1977). Mendy et al. (1979) found that whilst mature wheat grains showed pyrrolooxygenase activity throughout, inhibition of the enzyme was located mainly in the embryo. Hagima et al. (1978) reported that characteristic varietal patterns of peroxidases and catalases from wheat endosperm became established after breaking of dormancy and persisted during germination. Mehta & Chinog (1978) found a relationship between ascorbic acid metabolism and the activity of peroxidases and catalase during the first 96 h of germination of *Plantago* seeds.

Phosphorus metabolism

Vorob'ev (1976) reported that the level of phosphorus compounds in rice embryo corresponded well with the germination rate. Mobilization of the aleurone phytate store did not begin until after 1 d of germination. In germinating soybean and peas, Chen & Pan (1977) found that more than half of the phytate still remained after five days. After 10 d germination 50 % of the original phytate content was found in *Phaseolus mungo* seeds (Reddy et al., 1978). In contrast, phytate disappeared completely from fenugreek seeds after 8 d of germination although there was little change during the first 48 h (Tayal et al., 1978).
A Co^{2+}-requiring phytase was purified from navy beans cv. Sanilac by Lolas & Markais (1977). The enzyme increased in activity during germination as the phytate store declined. Multiple forms of phytase have been purified from the cotyledons of germinating *Cucurbita maxima* by Goel & Sharma (1979). Papageorgakopoulou & Georgatos (1977) demonstrated that the number of acid phosphatases in barley increased from 7 to 25 after 120 h of germination. In an investigation on germinating tall and dwarf pea varieties, Hota & Israelstam (1978) could find no alkaline phosphatase activity, but showed a marked increase in acid phosphatase activity during inhibition. The acid phosphatases of *Agrostemma githago* L. increased sharply in activity during germination (Borriss & Schmerder, 1978). The same response was observed in *Allium cepa* by Colson (1979).

Etiolated cotyledons of pumpkin were found to lose almost 70 % of their protein content during 14 d germination (Hara et al., 1976). The globulins were rapidly degraded to soluble proteins, but the soluble fraction still maintained the oxaloacetate decarboxylase activity found originally in the globulin fraction. Reilly et al. (1978) found a much more rapid breakdown of the globulin of pumpkin seeds during germination. However, the molecular weights of the breakdown products appear to be very similar in both studies. Kumar & Vankataraman (1978) observed a stepwise degradation of the three main storage proteins of chickpea during a 12-d germination period with a sedimentation time falling from 1.03 ps to 0.88 ps, from 0.69 to 0.56 ps and from 0.22 to 0.10 ps. Konopska (1978) found a differential loss of protein fractions from the protein bodies of pea during germination. The legumin:vicilin ratio rose during the first 3 d but there was little change in the amounts of glutelins or prolamins. After 3 d (Knopska, 1979) vicilin continued to be utilized more rapidly than legumin but after 4 d this was reversed. Germination of dehulled glandless cottonseeds for 5 d did not lead to any change in the relative concentrations of the proteins (sedimentation time 0.2, 0.7 and 1.2 ps) although the acid soluble protein did increase (Cunningham et al., 1978). Khvostova (1979) showed that during the germination of castor beans, the quantity of the toxic glycoprotein, ricin, remained relatively unchanged over the first 5 d despite a pronounced decline in total protein.

A number of papers have appeared which are concerned with the isolation of protein bodies from seeds and the changes which occur in these organelles during germination: Kirk & Pyliotis (1976), *Sinapis alba*; Tully & Beevers (1976), castor bean; Youle & Huang (1976), castor bean; Pusztai et al. (1977), *Phaseolus vulgaris* L.; Pernollet (1978), a general review; Mlodzianowski (1978), *Lupinus luteus* and Weber et al. (1979), *Vicia faba*.

Tully & Beevers (1978) found that the thiol-dependent aminopeptidases present in the endosperm of castor beans increased 4-fold in activity to a maximum level on the fourth day of germination. At days 2 to 3, haemoglobin protease, carboxypeptidase and BANA-ase activities appeared and these reached a peak at around day 5 to 6. In wheat, Preston et al. (1978) found that although both exo- and endo-proteolytic activities were present initially, it was the latter which showed distinct rises during germination. However, the main product of proteolysis appeared to be amino acids implying that the exo-proteolytic enzymes have a significant activity within the endosperm. Moureaux (1979) demonstrated the presence of three acid proteases in the endosperm of germinating maize. One was present in ungerminated corn and could not hydrolyze prolamines, whereas the appearance of the other two enzymes coincided with the rapid breakdown of the protein fractions at day 2 to 2.5. In the case of both buckwheat and rye, BAPA-ase activity was found in the endosperm and the embryo (Dunaevskii et al., 1978). The endosperm enzymes either declined or showed little change during germination, whereas the activity of the embryo enzymes changed markedly with very distinct peaks occurring. Kruger & Preston (1978) reported relatively little alteration in the activity of aminopeptidases during the germination of wheat. Collier & Murray (1977) identified two L-leucyl β-naphthylamidases in pea cotyledons and showed that their activity correlated better with general metabolic activity than with the phase of protein removal. Murray et al. (1979) described a complex pattern of changes in the activity of proteases and peptidases in the radicle and primary root of germinating peas, but Murray (1979) claimed that the cotyledons were the main source of nitrogen for the growing radicle. There was no evidence for large scale breakdown of the radicle proteins. During germination of *Vicia faba*, it was demonstrated that rapid breakdown of protein in the cotyledons and transfer of nitrogen to the embryo began on day 7 (Lichtenfeld et al., 1979). Peaks of acid endopeptidase occurred on days 3 and 9 whilst BAPA-ase activity began to increase from day 7. Metabolism of the reserve proteins of mung beans is accompanied by a considerable synthesis of endopeptidase in the protein bodies (Chrispeels, 1976). Bul'Maga & Shutov (1977) purified an endopeptidase from germinating vetch seeds which hydrolyzed native legumin and vicilin to peptides with 9 to 16 amino acid residues. Shepard & Moore (1978) purified an acid protease from germinating seeds of *Lupinus angustifolius*. In peanuts, rapid depletion of the protein reserves between days 4 and 10 of germination was associated with an increase in activity of an acid and a neutral protease (Basha & Cherry, 1978). Leucine aminopeptidase activity followed the general trend in proteolytic activity. The naphthylamidases of germinating Scots pine seeds represent only a small proportion of the total pepdidase activity, but the pattern of change observed suggests that they may be involved in mobilization of the reserve proteins (Salmia & Mikola, 1976). The pattern found is similar to that of barley and pea. Later studies on Scots pine (Salmia et al., 1978) have implicated acid and neutral proteinases in the breakdown of storage proteins. Pitel & Durzan (1976) found that the histones of several Pinaceae species compared closely with those from peas and that changes during germination were mainly in the Fl fraction. Grellet & Guitton (1976) showed an overall fall in the histone content of pea embryos during germination with the greatest change occurring in the Fl fraction. Yoshida & Sasaki (1977) failed to find any changes in the histone components during the germination of wheat, but did observe noticeable reductions in two non-histone proteins associated with the chromatin.

Nucleic acids

The ribosomal proteins from a number of species of *Pinus, Abies* and *Pseudotsuga* were found to be similar (Pitel & Durzan, 1977). Imbibition for 48 h resulted in the formation of polysomes in *Pinus banksiana*. Ferrer et al. (1979) isolated ribonucleoprotein particles of sedimentation time 2 ps from germinating radish seeds. These particles contained heterogenous ribonucleic acid (RNA) (1.8 and 0.4 ps) and gradually disappeared during germination. Dziegielewski & Pawelkiewicz (1977) found that tRNA from dry and germinated seeds of *Lupinus luteus* never lost more than three bases from the 3' terminus. Augustyniak & Pawelkiewicz (1978) were able

to distinguish five species of lysine tRNA in germinating lupin seeds (*Lupinus luteus*). On the germination of lupin seeds the level of tRNA amino acylation increased and that of defective tRNAs decreased (Dziegielewski et al., 1979). Adenosine 5'-triphosphate (ATP) availability seems more likely to be a controlling factor in tRNA metabolism than the level of activating enzymes. For castor beans, a range of RNA species achieved their maximum concentration at days 3 to 4 of germination and the maximum ability to incorporate adenine and uridine occurred on the second day (Roberts & Lord, 1979). Polyadenylated RNA capable of supporting in vitro protein synthesis was isolated from three-d-old castor bean seedlings by Bowden-Bonnett & Lord (1979). Takaiwa & Tanifugi (1979) also found active synthesis of a range of RNA species by pea embryo axes during the 6th to 64th hours of germination. The total amount of RNA, and particularly rRNA, was shown to be increased during the germination of tomato seeds by an osmotic presowing treatment (Coolbear & Grierson, 1979). In *Agrostemma githago*, Hecker et al. (1979) observed that the desoxyribonucleic acid (DNA) content of 70 % of the nuclei had doubled after 120 h of germination.
An enzyme which hydrolysed depurinated DNA but not native DNA to any extent was found in germinating peas by Wani & Hadi (1976). Wani & Hadi (1979) reported the partial purification of an endonuclease from germinating peas active on single stranded DNA. Shivakumar & Padayatty (1976) showed that 3'-ribonucleotidase reached a peak in germinating rice at 24 h and were able to purify the enzyme. Inhibitor studies suggested that the enzyme was synthesised from preformed RNA. In germinating pea axes three types of RNAase were found by Takaiwa & Tanifugi (1978). These enzymes paralleled the rise in RNA synthesis and one of them was associated with the endoplasmic reticulum. Mathew et al. (1978) showed that gibberellic acid (GA), indoleacetic acid (IAA) and ascorbic acid all stimulated RNAase production in dark-germinated barley although they had differential effects on other areas of RNA metabolism.

Physiologically active compounds

Tillberg (1977) found that the level of IAA in *Phaseolus vulgaris*, *Zea mays* and *Pinus sylvestris* was species-specific and influenced by the length of imbibition up to a maximum. Ethylene production by cotton, maize and broad bean was also species-specific, but declined throughout germination (Afifi et al., 1977). The G2 arrest factor, N-methylnicotinic acid, in pea cotyledons is transported to the root meristems during germination, where it causes cell arrest in cells that undergo normal cell differentiation (Evans, 1979). Von Schirach-Szmigiel (1979) showed that the quantity of gibberellin-like substances in *Phaseolus vulgaris* declined steadily over 3 d of germination.
Orf et al. (1977) reported that although there was no change in the PAGE behaviour of the Kunitz trypsin inhibitor proteins ($SBTl\text{-}A_2$) of soybeans during the first 3 d of germination, between days 4 and 6 some of the original bands disappeared and were replaced by new bands which retained trypsin inhibitor activity. Freed & Ryan (1978) used immunochemical techniques to confirm the earlier observation

of Orf et al. (1977). The trypsin inhibitor of *Vicia faba* was found by Bhatty (1977) to increase to a peak on day 3 and 4 of germination. The inhibitor was mainly present in the cotyledons. Xavier-Filho & Newton do Monte Negreiros (1979) found that initially both trypsin and chymotrypsin inhibitor activities were higher in the axis than the cotyledon of *Vigna sinensis*. After 24 h imbibition the situation was reversed and thereafter a decline in inhibitor activity occurred in the cotyledon as well as the axis.

Sal'nikov (1975) showed that, in wheat, germination at low temperature in over-moist soil caused a reduction in the amount of nicotinic acid, thiamine, inositol and pyridoxine in the seedling and an increase in biotin and panthothenic acid. Venugopal & Rao (1978) found that the vitamin C content of blackgram reached a pronounced maximum at 2 d of germination.

Intermediary metabolism

During the germination of castor bean, pyruvate kinase and phosphofructokinase showed a steady rise in activity in the endosperm until day 5 when a decline began (Nakayama et al., 1978). In *Cicer arietinum* glycolytic intermediates reached a maximum concentration in the endosperm after 1 d and after 2 d in the embryonic axis (Aldasoro & Nicolas, 1979). The results suggested a regulatory role for phosphofructokinase and pyruvate kinase. In the same organism, Nicolas & Aldasoro (1979) found that despite a rise in the activity of some of the enzymes of the pentose phosphate pathway the relative evolution of C_6/C_1 of glucose suggested that glycolysis was the major pathway. Ashihara & Matsumura (1977) showed that the pentose pathway was active in germinating seeds of *Phaseolus mungo* but could not estimate its importance because of substantial alcoholic fermentation.

Pyruvate kinase increased rapidly in the endosperm of castor beans after day 3 of germination (Nakayama et al., 1976). A partial purification of the enzyme was achieved. Seven-day-old seedlings of maize responded to anaerobiosis first by production of lactate and then ethanol (Leblová et al., 1976). An alcohol dehydrogenase was isolated and its properties studied. Vorojev (1977) showed that alcohol dehydrogenase was very active in rice just before the emergence of the radicle. Leblová & Perglerová (1977) isolated an alcohol dehydrogenase from germinating cucumber seeds. Crawford (1977) has suggested that in all seeds limitation of ethanol production is the key to survival in low-oxygen environments.

The microbody malate dehydrogenase in the cotyledons of *Cucumis sativus* underwent continuous turnover during germination, but its properties did not change even during greening (Wainwright & Ting, 1976). Fritsch & Beevers (1979) showed that ATP citrate lyase was at a maximum in the plastids of castor beans after 4 to 5 d of germination. Several dehydrogenases from the cotyledons of germinating soybeans showed breaks in their Arrhenius plots between 6 and 18 °C (Duke et al., 1977). The plots for nicotinamide dinucleotide phosphate (NADP)-isocitrate dehydrogenase were very similar to those for total mitochondrial respiration. Babenko et al. (1976) found no effect of low temperature on several dehydrogenases from spring and winter wheats.

Fuente Burguillo & Nicolas (1977) found that cyanide resistance of respiration in *Cicer arietinum* was at a maximum between 3 and 4 d of germination. The acetyl CoA carboxylase of plastids in sunflower seeds declined steadily during germination in the dark (Bhatia et al., 1978).

Ashihara (1977) showed that de novo synthesis of pyrimidine nucleotides was important in the embryonic axis of germinating black gram. Uracil salvage occurred only in the cotyledons whilst uridine salvage was active in both cotyledons and axis. Adenosylhomocysteinase and adenosine nucleosidase activities reached maximum levels in *Lupinus luteus* cotyledons at 4 to 5 d of germination (Guranowski & Pawelkiewicz, 1978). All parts of the germinating wheat seed were found to contain nucleotide pyrophosphatase activity by Gaham et al. (1979). The endosperm cells under the aleurone and the vascular tissues were especially active. Hoffman (1979) found that the smaller the reserve of adenylate, the sooner de novo synthesis of adenylates began on germination. With germinating *Sorghum vulgare*, Bhatia & Uppal (1979) found an initial decline in nucleotides followed by a rise after about 2 d.

During the germination of cottonseed the cotyledon amino acid pool expands considerably particularly with regard to asparagine (Capdecilla & Dure, 1977). This accumulation appears to be regulated by asparagine synthetase (Dilworth & Dure, 1978). Analysis of the amino acids in germinating cotton by Elmore & King (1978) also identified asparagine as the predominant compound. Two methionine based methyl transferase systems were studied during the germination of winter wheat by Allamong & Abrahamson (1977) and considerable changes found. Ogawa et al. (1978) found that a number of D-alanine derivatives increased rapidly in the early stages of pea seed germination. Synthesis appears to occur by racemization. The incorporation of proline into proteins during the germination of barley was affected by temperature; at 2 °C incorporation was higher in the cationic fraction and lower in the acid-neutral fraction than at 25 °C (Shiomi & Hori, 1978).

Diamine oxidase was detected in both cotyledons and embryo of germinating peas by Srivastava et al. (1977) by 62 h and appears to be induced by putrescine, spermidine and ornithine. In germinating *Phaseolus mungo* seeds, Villaneuva et al. (1978) detected putrescine, spermidine, spermine, cadaverine, agmatine and tyramine. The rate of biosynthesis was at a maximum 3 h from the beginning of germination. The polyamine content of *Pisum sativum, Tragopogon porrifolius, Zea mays* and *Triticum aestivum* was also investigated during the first 12 h of germination. Sindhu & Desai (1980) showed that the activity of agmatine iminohydrolase in the embryo of germinating peanuts depends on the presence of at least one cotyledon.

Higgins & Payne have published three papers concerned with the uptake of peptides by germinating isolated barley embryos. Higgins & Payne (1978a) showed that peptides up to pentaalanine were absorbed and that the rate was greater than for amino acids. Higgins & Payne (1978b) described competition experiments which indicated that a single peptide transport system occurs and this seems to be the same as the glycylsarcosine uptake system. The system favours peptides containing L-amino acid residues (Higgins & Payne, 1978c).

Initiation of active metabolism

Kedzierski & Pawelkiewicz (1977) showed that the level of aminoacylation of tRNA species was unaltered 1 h after the beginning of germination of *Lupinus luteus* seeds. Significant increase had occurred after 3 h. Bray & Dasgupta (1976) demonstrated RNA and protein synthesis in the embryonic axis of *Pisum arvense* winthin 1 h of germination beginning. Loss of viability went along with loss of RNA synthesising ability. Isolated wheat embryos began to make rRNA within the first 3 h of germination whilst in non-isolated embryos 6-12 h elapsed before this was detected (Grzelczak & Buchowicz, 1977). When seeds of *Agrostemma githago* were soaked with a sufficient supply of oxygen, synthesis of heterogenous RNA had begun within 30 min and rRNA by 1 h. tRNA and poly(A)-containing RNA were also formed very early. With the same plant, Hecker (1977) showed that most of the poly(A) sequences were associated with newly synthesised mRNA but that some polyadenylation of conserved mRNA also occurred. Payne (1977) claimed that rye embryos begin to synthesise poly A-rich RNA as soon as they begin to imbibe water. Payne et al. (1978) showed that with rape seed, synthesis of rRNA, tRNA and poly(A)-rich RNA began in the peripheral cells of the embryo near the testa soon after the uptake of water had begun. After 1 h of germination Iskhakov et al. (1978) could isolate polyribosomes from isolated wheat embryos and liberate particles containing heterogenous poly(A)-rich complexes of sedimentation time 0.8-1.5 ps. As well as RNA these complexes contained two proteins. Caers et al. (1979) showed that in the embryos of wheat seeds germinated for 6 h, more than half the mRNA was actively involved in translation. No preference was found for preformed or newly synthesised poly(A)-rich RNA. Hecker & Kohler (1979) reported that synthesis of protein and RNA starts right from the beginning of imbibition of *Vaccaria pyramidata* seeds.

Hecker et al. (1977) reported that the process of DNA synthesis was activated early in the after-ripening of *Agrostemma githago* seeds and was detectable before any increase in germination rate beame apparent. Fukuei et al. (1977) felt that the results they obtained on the effect of actinomycin D on freshly germinating *Vicia faba* embryos implied that synthesis of heterodisperase RNA is necessary for the synthesis of DNA. In wheat embryos, DNA synthesis starts about 12 h after the beginning of imbibition (Castroviejo et al., 1979). DNA polymerases A and B increased to a maximum level of activity at 15 h, whilst DNA polymerase C declined from the beginning. Khan & Upadhyaya (1979) found that in germinating barley embryos synthesis of nuclear DNA began by the 10th hour but synthesis of mitochondrial DNA did not begin until 15-17 h when the S phase was just past its peak

A thesis (Subramanian, 1972) and a review (Hecker, 1978) concerned with synthesis of protein and nucleic acids early in the germination of seeds have been published.

By use of ^{14}C-leucine incorporation experiments, Sobolev (1976) showed that synthesis of storage protein occurred in the aleurone grains of castor beans between 7 and 19 h of soaking presumably using residual mRNA from the developing seed stage. Fukushi et al.

(1977) found a sharp peak in the ^{3}H-leucine incorporating activity of wheat embryos after 36 h imbibition. Dry embryos of *Pinus thunbergii* seeds contain active ribosomes (Yamamoto, 1977), but no increase in their activity occurred during imbibition in the dark. There was an increase in the activity of the fraction separated in a centrifugal field of 0.98 Mm/s^2 but both samples depended on addition of exogenous mRNA. When germinated in the light, the embryos contained polysomes and extracts could support protein synthesis without addition of mRNA (Yamamoto & Sasaki, 1977). Yamamoto & Sasaki (1979) showed that the dissociation of the ribosomes from dry pine was abnormal. The change to normal ribosome characteristics which occurred during imbibition could also be observed in vitro when the separated ribosomes were stored in salt solutions.
Alexandrescu et al. (1976) showed that distinct changes in the electrophoretic patterns of catalase, peroxidase, ribonuclease, acid phosphatase, esterase and amylases occurred during stratification of seeds of *Tilia platyphyllos*. They suggested that these changes are necessary for germination to occur. Gaspar et al. (1977) suggested that a factor in wheat germination is a fall in peroxidase activity below a critical level. The acid protease and phosphatase activities of the aleurone grains from pea seeds fell steadily from the first hour of imbibition (Konopska & Sąkowski, 1978). Colson (1979) used a lead staining technique to visualize acid phosphatase activity during the germination of *Allium cepa* seeds. After 3 h of imbibition the activity increased in the aleurone grains and began to appear in other parts of the cells.
Stevens & Stevens (1977) described a technique for measurement of glucose-6-phosphate dehydrogenase activity at low water concentrations and used it to estimate the activity of this enzyme in resting barley embryos.
Anderson (1979) found that isolated embryonic axes of soybean could convert ^{14}C-glycine to ATP after 1 h of imbibition and to guanosine 5'-triphosphate (GTP) after 3 h. Morohashi (1978) showed that during the early phase of imbibition of *Phaseolus mungo* seeds the main amino acid synthesised from U-^{14}C-glucose was alanine. Aspartic acid concentration declined during this time and an active aspartate: pyruvate aminotransferase was detected. In wheat, an 8-fold increase in the uridine 5'-diphosphate (UDP)-glucose content of isolated embryos occurred during the first 40 min of imbibition (Cheung & Suhadolnik, 1979).
Using both the lettuce hypocotyl and barley endosperm tests, Magy & Gonder (1977) were able to show a very large increase in concentration of gibberellin-like substances of *Lupinus alba* during the first few hours of water uptake. Cocucci & Cocucci (1977) observed a release of K^+, Na^+ Mg^{2+} and Ca^{2+} during the first 10 h of germination of *Rhaphanus sativus* seeds. This was followed by a period of active K^+ and Na^+ uptake. The germination stimulator, fusicoccin, accelerated the rate of cation uptake.

References

Afifi, A.F., M.F. Ebid & A.A. El-Gindy, 1977. Ethylene production during seed germination of some plants. Zentralblatt für Bakteriologie, Parasitenkunde, Infektionskrankheiten und Hygiene 132(8): 673-676.

Aldasoro, J.J. & G. Nicolás, 1979. Changes in the concentrations of glycolytic intermediates and adenosine phosphates during germination of seeds of Cicer arietinum L. The International Journal of Biochemistry 10(11): 947-950.

Alexandrescu, V., I. Hagima, D. Popov, F. Nihailescu & Z. Dobrescu, 1976. Multiple molecular forms of some enzymes during the grain dormancy in Tilia platyphyllos. Revue Roumaine de Biochimie 13(3): 163-167.

Allamong, B.D. & L. Abrahamson, 1977. Methyltransferase activity in dry and germinating wheat seedlings. Botanical Gazette 138(1): 46-51.

Allfrey, J.M. & D.H. Northcote, 1977. Rhe effects of the axis and plant hormones on the mobilization of storage materials in the ground-nut (Arachis hypogaea) during germination. New Phytologist 78(3): 547-563.

Åman, P., 1979. Carbohydrates in raw and germinated seeds from mung bean and chick pea. Journal of the Science of Food and Agriculture 30(9): 869-875.

Amuti, K.S. & C.J. Pollard, 1977. The metabolism of galactose and the raffinose oligosaccharides by germinating bambarra groundnut seeds. Phytochemistry 15(5): 533-537.

Anderson, J.D., 1979. Purine nucleotide metabolism of germinating soybean embryonic axes. Plant Physiology 63(1): 100-104.

Ashihara, H., 1977. Changes in activities of the de novo and salvage pathways of pyrimidine nucleotide biosynthesis during germination of black gram (Phaseolus mungo) seeds. Zeitschrift für Pflanzenphysiologie 81(3): 199-211.

Ashihara, H. & H. Matsumura, 1977. Changes in the activity and the function of the pentose phosphate pathway during germination of black gram (Phaseolus mungo) seeds. The International Journal of Biochemistry 8(6): 461-471.

Augustyniak, H. & J. Pawełkiewicz, 1978. Lysyl tRNase of Lupinus luteus seeds. Phytochemistry 17(1): 15-18.

Babenko, V.I., M.L. Makhnovskaya & I.P. Ruzhitskaya, 1976. Influence of low temperature on the activity of certain dehydrogenases in wheat seeds. Soviet Agriculture Sciences 5: 1-3.

Bacic, A., G.N. Richards & D. Yellowlees, 1977. Polysaccharase metabolism in dormant and germinating Stylosanthes humilis seeds. Phytochemistry 16(7): 845-848.

Ballance, G.M., W.O.S. Meredith & D.E. LaBerge, 1976. Distribution and development of endo- β-glucanase activities in barley tissues during germination. Canadian Journal of Plant Science 56(3): 459-466.

Basha, S.M. & J.P. Cherry, 1978. Proteolytic enzyme activity and storage protein degradation in cotyledons of germinating peanut (Arachis hypogaea L.) seeds. Journal of Agricultural and Food Chemistry 26(1): 229-234.

Bhatia, I.S., K.L. Ahuja & P.S. Sukhija, 1978. Changes in the activity of acetyl CoA carboxylase in germinating and ripening sunflower seeds. Physiologia Plantarum 44(3): 141-144.
Bhatia, I.A., K.K. Arora & P.S. Sukhija, 1978. Changes of lipids and lipase activity in the germinating sunflower seeds. The Plant Biochemical Journal 5(2): 150-156.
Bhatia, I.S. & D. Uppal, 1979. Free and sugar nucleotides in seedlings of Sorghum vulgare. Plant Science Letters 16(1): 59-66.
Bhatty, R.A., 1977. Trypsin inhibitor activity in faba beans (Vicia faba var. minor), changes during germination and distribution. Canadian Journal of Plant Science 57(3): 979-982.
Borriss, H. & B. Schmerder, 1978. Hormonale regulation der phosphatase-aktivität in nachgereiften und dormanten embryonen von Agrostemma githago L. Biochemie un Physiologie der Pflanzen 172(5): 453-474.
Bouquelet, S. & G. Spik, 1978. Properties of four molecular forms of N-acetyl-β-D-Hexosaminidase isolated from germinating seeds of fenugreek (Trigonella foenum graecum). European Journal of Biochemistry 84(2): 551-559.
Bowden-Bonnett, L. & J.M. Lord, 1979. Isolation and cell-free translation of total messenger RNA from germinating castor bean endosperm. Plant Physiology 63(4): 769-773.
Bray, C.M. & J. Dasgupta, 1976. Ribonucleic acid synthesis and loss of viability in pea seed. Planta (Berl.) 132(2): 103-108.
Bul'Maga V.P. & A.D. Shutov, 1977. Partial purification and characterization of protease a of germinating vetch seeds, hydrolyzing native reserve proteins. Biochemistry 42(11, part 1): 1563-1568.
Caers, I., W.J. Peumans & A.R. Carlier, 1979. Preformed and newly synthesised messenger RNA in germinating wheat embryos. Planta 144(5): 491-496.
Capdevila, A.M. & L. Dure III, 1977. Developmental biochemistry of cottonseed embryogenesis and germination. VIII. Free amino acid pool composition during cotyledon development. Plant Physiology 59(2): 268-273.
Castroviejo, M., D. Tharaud, B. Mocquot & S. Litvak, 1979. Factors of affecting the onset of deoxyribonucleic acid synthesis during wheat embryo germination. The Biochemical Journal 181(1): 193-199.
Chen, L.H. & S.H. Pan, 1977. Decrease of phytates during germination of pea seeds (Pisum sativa). Nutrition Reports International 16(1): 125-131.
Cheung, C.P. & R.J. Suhadolnik, 1979. Rapid increase in UDP-glucose during early wheat embryo germination. Plant Physiology 63(1): 146-148.
Chrispeels, M.J., 1976. Regulation of reserve protein metabolism in the cotyledons of mung bean seedlings. Proceedings of the National Academy of Sciences 73(9): 3168-3172.
Cocucci, S. & M. Cocucci, 1977. Effect of ABA, GA_3 and FC on the development of potassium uptake in germinating radish seeds. Plant Science Letters 10(1): 85-95.
Collier, M.D. & D.R. Murray, 1977. Leucil β-naphthyl-amidase activities in developing seeds and seedlings of Pisum sativum L. Australian Journal of Plant Physiology 4(4): 571-582.

Colson, M., 1979. Localisation ultrastructurale de l'activité phosphatasique acide et des phosphates dans les radicules d'Allium cepa en vie ralentie et au début de l'hydratation germanitive. Effet de traitement brefs à basse température (-196 °C). Botanique et Biologie Végétale Annales de Science Naturelles 13(1): 21-38.

Coolbear, P. & D. Grierson, 1979. Studies on the changes in the major nucleic acid components of tomato seeds (Lycopersicon esculentum Mill.) resulting from osmotic presowing treatment. Journal of Experimental Botany 30(119): 1153-1162.

Crawford, R.M.M., 1977. Tolerance of anoxia and ethanol metabolism in germinating seeds. The New Phytologist 79(3): 511-517.

Cunningham, S.D., C.M. Cater & K.F. Mattil, 1978. Effects of germination on cottonseed protein. Journal of Food Science 43(1): 102-105.

Daussant, J. & R.D. Hill, 1979. Immunochemical identification of α-amylases in developing, mature and germinated Triticale seeds. Physiologia Plantarum 45(2): 255-259.

Davis, B.D., 1977. Occurrance of a-Amylase in the axis of germinating peas. Plant Physiology 60(4): 513-517.

Dendsay, J.P.S. & R.C. Sacher, 1978. Hormonal control of peroxidase activity in germinating mung bean cotyledons. Phytochemistry 17(6): 1017-1019.

Dilworth, M.F. & L. Dure III, 1978. Developmental biochemistry of cotton seed embryogenesis and germination. X. Nitrogen flow from arginine to asparagine in germination. Plant Physiology 61(4): 698-702.

Dogras, C.C., D.R. Dilley & R.C. Herner, 1977. Phospholipid biosynthesis and fatty acid content in relation to chilling injury during germination of seeds. Plant Physiology 60(6): 897-902.

Doireau, P., 1977. Les amyrines des semences de pois évolution de ces substances au cours de la germination. Physiologie Végétale 16(2): 291-299.

Donaldson, R.P. & H. Beevers, 1978. Organelle membranes from germinating castor bean endosperm. I. A comparison of purified endoplasmic reticulum, glyoxysomes and mitochondria. Protoplasma 97(4)- 317-327.

Duke, S.H., L.E. Schrader & M.G. Miller, 1977. Low temperature effects on soybean (Glycine max (L.) Merr cv. Wells) mitochondrial respiration and several dehydrogenases during imbibition and germination. Plant Physiology 60(5): 716-722.

Dunaevskii, Y.E., I.B. Emtseva & M.A. Belozerskii, 1978. Localization and variation of BAPAase activity during the germination of buckwheat and rye seeds. Moscow University Biological Sciences Bulletin 33(1): 60-62.

Dupéron, P., J. Meance & R. Dupéron, 1979. Métabolisme des différentes catégories de substances stéroliques au cours de la germination et pendant la croissance des jeunes plantes de Laitue (Lactuca sativa L.). Physiologie Végetale 17(4): 839-850.

Dziegielewski, T., W. Kedzierski & J. Dawelkiewicz, 1979. Levels of aminoacyl-tRNA synthetases, tRNA nucleotidyltransferase and ATP in germinating lupin. Biochimica et Biophysica Acta (N) Nucleic Acids and Protein Synthesis 564(1): 37-42.

Dziegielewski, T. & J. Pawelkiewicz, 1977. Changes of tRNA CCA end

during germination of Lupinus luteus seeds. Bulletin de l'Académie Polonaise des Sciences; Série des Sciences Biologiques 25(7): 433-435.
Elmore, C.D. & E.E. King, 1978. Amino acid composition of germinating cotton seeds. Plant Physiology 62(4): 531-535.
Evans, L.S., 1979. Developmental aspects of roots of Pisum sativum influenced by the G2 factor from cotyledons. American Journal of Botany 66(8): 880-886.
Ferrer, A., M. Delseny & Y. Guitton, 1979. Isolation and characterisation of subribosomal ribonucleoprotein particles from radish seeds and seedlings. Plant Science Letters 14(1): 31-42.
Freed, R.C. & D.S. Ryan, 1978. Note on modification of the kunitz soybean trypsin inhibitor during seed germination. Cereal Chemistry 55(4): 534-538.
Fritsch, H. & H. Beevers, 1979. ATP citrate lyase from germinating cator bean endosperm. Plant Physiology 63(4): 687-691.
Fuente Burguillo, P. de la & G. Nicolás, 1977. Appearance of an alternate pathway cyanide-resistant during germination of seeds of Cicer arietinum. Plant Physiology 60(4): 524-527.
Fukuei, K., T. Sakamaki, N. Takahashi, F. Takaiwa & S. Tanifuji, 1977. RNA synthesis required for DNA replication in Vicia seed embryos. Plant and Cell Physiology 18(1): 173-180.
Fukushi, S., K. Ishikawa, & K. Sasaki, 1977. In vitro protein synthesis during germination and vernalization in winter wheat embryos. Plant and Cell Physiology 18(5): 969-977.
Gahan, P.B., H. Sierakowska & A.L. Dawson, 1979. Nucleotide pyrophosphatase activity in dry germinated seeds of Triticum and its relationship to general acid phosphatase activity. Planta 145(2): 159-166.
Gaspar, T., R. Wyndaele, M. Bouchet & E. Ceulemans, 1977. Peroxidase and α-amylase activities in relation to germination of dormant and nondormant wheat. Physiologia Plantarum 40(1): 11-14.
Goel, M. & C.B. Sharma, 1979. Multiple forms of phytase in germinating cotyledons of Cucurbita maxima. Phytochemistry 18(12): 1939-1942.
Goswami, A.K., M.K. Jain & B. Paul, 1977. α- and β-Amylases in seed germination. Biologia Plantarum 19(6): 469-471.
Grellet, F. & Y. Guitton, 1976. Métabolisme des histones au cour des premières heures du développement de l'embryon de pois (Pisum sativum L.). Physiologie Végétale 14(3): 453-465.
Grzelczak, Z. & J. Buchowicz, 1977. A comparison of the activation of ribosomal RNA synthesis during germination of isolated and non-isolated embryos of Triticum aestivum L. Planta 134(3): 263-265.
Guranowski, A. & J. Pawełkiewicz, 1978. Adenosylhomocysteinase and adenosine nucleosidase activities in Lupinus luteus cotyledons during seed formation and germination. Planta 139(3): 245-247.
Hagima, I., V. Alexandrescu & Z. Cseresnyes, 1978. Peroxidases and catalases of dormant, able to germinate and germinated wheat seeds. Revue Roumaine de Biochimie 15(4): 273-277.
Halmer, P. & J.D. Bewley, 1979. Mannanase production by the lettuce endosperm. Control by the embryo. Planta 144(4): 333-340.
Halmer, P., J.D. Bewley & T.A. Thorpe, 1978. Degradation of the

endosperm cell walls of Lactuca sativa L., cv. Grand Rapids. Timing of mobilisation of soluble sugars, lipid and phytate. Planta 139(1): 1-8.

Hara, I., K. Wada, S. Wakabayashi & H. Matsubara, 1976. Pumpkin (Cucurbita sp.) seed globulin. II. Alterations during germination. Plant and Cell Physiology 17(4): 815-823.

Hecker, M., 1977. Gibt es eine Polyadenylation langlebigen Embryo-mRNA in den frühen Quellungsstadien von Agrostemma-Samen? Biologische Rundschau 15(1): 58-61.

Hecker, M., 1978. Nucleinsüren- und Proteinsynthesen in quellenden pflanzlichen Embryonen und ihre Bedeutung für den Aktivitätswechsel von Samen. Biologische Rundschau 16(3): 137-154.

Hecker, M., D. Bernhardt & H. Dunger, 1977. DNA-Synthese während der Nachreife von Agrostemma githago-Samen. Biologische Rundschau 15(5): 308-311.

Hecker, M., D. Bernhardt, K.H. Köhler, B. Schmerder, H. Dunger & R. Knösche, 1979. Synthesis of DNA and of hydrolytic enzymes in cotyledons of germinating seeds of Agrostemma githago L. Biochemie und Physiologie der Pflanzen 174(7): 523-536.

Hecker, M. & K.H. Köhler, 1979. Protein and nucleic acid synthesis during imbibition of dormant and afterripened Vaccaria pyramidata seeds. Developmental Biology 69(1): 270-280.

Hecker, M., K.H. Köhler & M. Wiedmann, 1977. Reactivation of Ribonucleic acid synthesis during early germination of Agrostemma githago embryos. Biochemie und Physiologie der Pflanzen 171(5): 401-408.

Higgins, C.F. & J.W. Payne, 1978a. Peptide transport by germinating barley embryos: uptake of physiological Di- and Oligopeptides. Planta 138(3): 211-215.

Higgins, C.F. & J.W. Payne, 1978b. Peptide transport by germinating barley embryos: evidence for a single common carrier for D- and Oligopeptides. Planta 138(3): 217-221.

Higgins, C.F. & J.W. Payne, 1978c. Stereospecificity of peptide transport by germinating barley embryos. Planta 142(3): 299-305.

Hizukuri, S. & N. Nakamura, 1978. Studies on plant starch granule-degrading enzyme. I. Activation of Azuki seed (Phaseolus angularis W.F. Wight) enzyme by inorganic phosphate. Bulletin of the Faculty of Agriculture Kagoshima University (28): 131-137.

Hoffmann, P., 1979. Der Adenylatgehalt in lufttrockenen und keimenden Karyopsen ausgewähter Weizen-Evolutionsformen. Biochemie und Physiologie der Pflanzen 174(3): 235-239.

Hota, C. & G.F. Israelstam, 1978. Phosphatase activity and gibberellin levels in seeds of dwarf and normal cultivars of pea (Pisum sativum L.). Zeitschrift für Pflanzenphysiologie 86(3): 241-249.

Huang, A.H. & R.A. Moreau, 1978. Lipases in the storage tissues of peanut and other oil seed during germination. Planta 141(1): 111-116.

Igaue, I., K. Kumagai, R. Togawa, I. Nakajima, F. Kurasawa & T. Hayakawa, 1978. Isolation and properties of α-amylase from germinated rice seeds. Bulletin of the Faculty of Agriculture Niigata University (30): 63-76.

Iskhakov, B.K., N.G. Filimonov & M.A. Ajtkhozhin, 1978. Proteins bound to Poly(A) sequences of polyribosomes from germinating

wheat embryos. Biochimica et Biophysica Acta(N) Nucleic Acids and Protein Synthesis 521(2): 470-475.
Jacobson, A. & M.R. Corcoran, 1977. Tannins as gibberellin antagonists in the synthesis of α-amylase and acid phosphatase by barley seeds. Plant Physiology 59(2): 129-133.
Kedzierski W. & J. Pawełkiewicz, 1977. Effect of seed germination on levels of tRNA aminoacylation. Phytochemistry 16(5): 503-504.
Khan, M.A. & K.C. Upahyaya, 1979. Synthesis of mitochondrial DNA in relation to nuclear cell cycle in germinating barley embryos. Zeitschrift für Pflanzenphysiologie 94(2): 159-161.
Khvostova, I.V., 1979. Changes in composition of the protein complex of castor seeds during ripening and germination. Soviet Plant Physiology 26(3): 422-426.
Kirk, J.T.O. & N.A. Pyliotis, 1976. Cruciferous oilseed proteins: the protein bodies of Sinapis alba seed. Australian Journal of Plant Physiology 3(6): 731-746.
Konopska, L., 1978. Proteins of aleurone grains in germinating seeds of Pisum sativum L. Biochemie und Physiologie der Pflanzen 173(4): 322-326.
Konopska, L., 1979. Changes in protein composition of aleurone grains of Pisum sativum L. during germination. Biochemie und Physiologie der Pflanzen 174(4): 275-282.
Konopska, L. & R. Sakowski, 1978. Enzyme activities in aleurone grains of pea seeds during the early stages of germination. Biochemie und Physiologie der Pflanzen 173(6): 536-540.
Krause, J., 1978. Über die Bedeutung des Endosperms für die Flavonoidsynthese in den Kotyledonen von Fagopyrum esculentum Moench. Zeitschrift für Pflanzenphysiologie 90(4): 319-330.
Krüger, J.E. & K.R. Preston, 1978. Changes in aminopeptidases of wheat kernels during growth and maturation. Cereal Chemistry 55(3): 360-373.
Kumar, K.G. & L.V. Venkataraman, 1978. Chickpea seed proteins: modification during germination. Phytochemistry 17(4): 605-609.
Leblová, S. & E. Perglerová, 1977. Alcohol dehydrogenase from cucumber seeds. Biochemie und Physiologie der Pflanzen 171(1): 1-6.
Leblová, S., J. Zima & E. Perglerová, 1976. Conversion of pyruvate under natural and artificial anaerobiosis in maize. Australian Journal of Plant Physiologie 3(6): 755-761.
Leung, D.W.M., J.S.G. Reid & J.D. Bewley, 1979. Degradation of the endosperm cell walls of Lactuca sativa L., cv. Grand Rapids in relation to the mobilisation of proteins and the production of hydropytic enzymes in the axis, cotyledons and endosperm. Planta 146(3): 335-341.
Lichtenfeld, C, R. Manteuffel, K. Müntz, D. Neumann, G. Scholz & E. Weber, 1979. Protein degradation and proteolytic activities in germinating field beans (Vicia faba L., var. minor). Biochemie und Physiologie der Pflanzen 174(4): 255-274.
Lockwood, G.B. & K.R. Brian, 1976. Influence of hormonal supplementation on steroid levels during callus induction from seeds of Trigonella foenumgraecum. Phytochemistry 15(11): 1655-1660.
Lolas, G.M. & P. Markakis, 1977. The phytase of navy beans (Phaseolus vulgaris). Journal of Food Science 42(4): 1094-1097.

Marbach, I., A.M. Mayer & R. Maron, 1978. Galactosidases in cultivated and wild peas. Phytochemistry 17(4): 655-657.

Matheson, N.K. & R.H. Richardson, 1978. Kinetic properties of two starch phosphorylases from pea seeds. Phytochemistry 17(2): 195-200.

Matheson, N.K. & H.S. Saini, 1977. Polysaccharide and oligosaccharide changes in germinating lupin cotyledons. Phytochemistry 16(1): 59-66.

Mathew, T., I.C. Dave & B.K. Gauer, 1978. Modification of RNA metabolism in barley by gibberellic acid, indole-3-acetic acid and 1-ascorbic acid. Zeitschrift für Pflanzenphysiologie 86(1): 23-29.

Maudinas, B. & J. Lematre, 1979. Violaxanthin, the major carotenoid pigment in Zea mays root cap during seed germination. Phytochemistry 18(11): 1815-1817.

Mehta, K.G. & J.J. Chinoy, 1978. Pattern of ascorbic acid metabolism during the germination phase of plantago (Plantago ovata Forsk) in relation to thermal pretreatment and the exogenous application of ascorbic acid and hydrogen peroxide. Indian Journal of Plant Physiology 21(2): 142-149.

Mendy, D.F., M.L. Tomaro & R.B. Frydman, 1979. Pyrrolooxygenases from Argentine wheat varieties. Phytochemistry 18(9): 1427-1431.

Minamikawa, T., 1977. Quinate:NAD oxidoreductase of germinating Phaseolus mungo seeds: Partial purifiction and some properties. Plant and Cell Physiology 18(4): 743-752.

Minamikawa, T. & T. Koshiba, 1979. Histochemical studies on mobilization of storage components in cotyledons of germinating Phaseolus mungo seeds. The Botanical Magazine Tokyo 92(1028): 325-332.

Młodzianowski, F., 1978. The fine structure of protein bodies in lupine cotyledons during the course of seed germination. Zeitschrift für Pflanzenphysiologie 86(1): 1-13.

Morohashi, Y., 1978. Some events of glucose catabolism occurring in the early imbibitional stage of germination of Phaseolus mungo seeds. International Journal of Plant Physiology 90(3): 209-214.

Morrall, P. & D.E. Briggs, 1978. Changes in cell wall polysaccharides of germinating barley grains. Phytochemistry 17(9): 1495-1502.

Moureaux, T., 1979. Protein breakdown and protease properties of germinating maize endosperm. Phytochemistry 18(7): 1113-1117.

Murray, D.R., 1979. Proteolysis in the axis of the germinating pea seed. II. Changes in polypeptide composition. Planta 147(2): 117-121.

Murray, D.R., M.B. Peoples & S.P. Waters, 1979. Proteolysis in the axis of the germinating pea seed. I. Changes in protein degrading enzyme activities of the radicle and primary root. Planta 147(2): 111-116.

Nagy, M. & E. Göndör, 1977. Change in the endogenous gibberellin content during swelling of Lupinus albus L. seeds. Acta Biologica, Nova series 23(1-4): 39-47.

Nakayama, H., M. Fujii & K. Miura, 1976. Partial purification and some regulatory properties of pyruvate kinase from germinating castor endosperm. Plant and Cell Physiology 17(4): 653-660.

Nakayama, H., M. Fujii & K. Miura, 1978. Subcellular localization of glycolytic key enzymes in germinating castor bean endosperm.

Plant and Cell Physiology 19(5): 777-784.
Nicolás, G. & J.J. Aldasoro, 1979. Activity of the phosphate pathway and changes in nicotinamide nucleotide content during germination of seeds of Cicer arietinum L. Journal of Experimental Botany 30(119): 1163-1170.
Ogawa, T., Y. Kawasaki & K. Sasaoka, 1978. De novo synthesis of d-alanine in germinating Pisum sativum seedlings. Phytochemistry 17(8): 1275-1276.
Okamoto, K. & T. Akazawa, 1978. Purification of α- and β-Amylase from endosperm tissues of germinating rice seeds. Agricultural and Biological Chemistry 42(7): 1379-1384.
Okamoto K. & T. Akazawa, 1979. Enzymic mechanisms of starch breakdown in germinating rice seeds. 7. Amylase formation in the epithelium. Plant Physiology 63(2): 336-340.
Okamoto, K. & T. Akazawa, 1980. Enzymic mechanism of starch breakdown in germinating rice seeds. Plant Physiology 65(1): 81-84.
Orf, J.H., D.W. Mies & T. Hymowitz, 1977. Qualitative changes of the Kunitz trypsin inhibitor in soybean seeds during germination as detected by electrophoresis. Botanical Gazette 138(3): 255-260.
Pal, A., D.S. Wagle & V.S. Sheorain, 1976. Effect of germination temperature on amylolytic and proteolytic activities of Bajra and barley during germination. Journal of Food Science and Technology 13(5): 253-254.
Papageorgakopoulou N. & J.G. Georgatos, 1977. The phosphohydrolytic activity of dormant and germinated barley seeds. The International Journal of Biochemistry 8(9): 611-618.
Payne, P.I., 1977. Synthesis of poly a-rich RNA in embryos of rye during imbibition and early germination. Phytochemistry 16(4): 431-434.
Payne, P.I., M. Dobrzanska, P.W. Barlow & M. Gordon, 1978. The synthesis of RNA in imbibing seed of rape (Brassica napus) prior to the onset of germination: a biochemical and cytological study. Journal of Experimental Botany 29(108): 77-88.
Pernollet, J.-C., 1978. Protein bodies of seeds: ultrastructure biochemistry, biosynthesis and degradation. Phytochemistry 17(9): 1473-1480.
Pitel, J.A. & D.J. Durzan, 1976. Nuclear proteins of dry and germinating conifer seeds. Proceedings of the Twelfth Lake States Forest Tree Improvement Conference, USDA Forest Service, General Technical Report NC-26, p. 147-157.
Pitel, J.A. & D.J. Durzan, 1977. Ribosomes and ribosomal proteins of dry and germinating conifer seeds. Phytochemistry 16(10): 1475-1480.
Preston, K.R., J.E. Dexter & J.E. Kruger, 1978. Relationship of exoproteolytic and endoproteolytic activity to storage protein hydrolysis in germinating durum and hard red spring wheat. Cereal Chemistry 55(6): 877-888.
Pusztai, A., R.R.D. Croy, G. Grant & W.B. Watt, 1977. Compartmentalization in the cotyledonary cells of Phaseolus vulgaris L. seeds: a differential sedimentation study. The New Phytologist 79(1): 61-71.
Reddy, N.R., C.V. Balakrishnan & D.K. Salunkhe, 1978. Phytate phosphorus and mineral changes during germination and cooking of

black gram (Phaseolus mungo) seeds. Journal of Food Science 43(2): 540-543.

Reid, G., C. Davies & H. Meier, 1977. Endo-β-mannanase, the leguminous aleurone layer and the storage galactomannan in germinating seeds of trigonella foenum-graecum L. Planta 133(3): 219-222.

Reilly, C.C., B.T. O'Kennedy, J.S. Titus & W.E. Splittstoesser, 1978. The solubilization and degradation of pumpkin seed globulin during germination. Plant and Cell Physiology 19(7): 1235-1246.

Roberts, L.M. & J.M. Lord, 1979. Ribonucleic acid synthesis in germinating castor bean endosperm. Journal of Experimental Botany 30(117): 739-749.

Rodaway, S.J. & H. Kende, 1978. No effect of 5-fluorouracil on the properties of purified α-amylase from half-seeds. Plant Physiology 61(1): 1-6.

Salmia, M.A. & J.J. Mikola, 1976. Localization and activity of naphthylamidases in germinating seeds of Scots pine, Pinus sylvestris. Physiologia Plantarum 38(2): 73-77.

Salmia, M.A., S.A. Nyman & J.J. Mikola, 1978. Characterization of the proteinases present in germinating seeds of scots pine, Pinus sylvestris. Physiologia Plantarum 42(2): 252-256.

Sal'Nikov, A.I., 1975. Biosynthesis of vitamins by germinating seeds under conditions of low temperature and an over-moistened soil. Applied Biochemistry and Microbiology 11(6): 765-767.

Schirach-Szmigiel, L. von, 1979. Alterations in endogenous levels of gibberellin-like substances during germination of Phaseolus vulgaris seeds. Physiologia Plantarum 46(1): 54-57.

Seiler, A., 1977. Galaktomannanabbau in keimenden Johannisbrotsamen (Ceratonia siliqua L.). Planta 134(3): 209-221.

Sengupta, T., B. Ghosh & S.M. Sircar, 1977. Effect of growth substances on the IAA oxidase-peroxidase activity during seed germination of rice (Oryza sativa L.). The Plant Biochemical Journal 4(1): 28-33.

Shepard, D. & K.G. Moore, 1978. Purification and properties of a protease from Lupinus angustifolius during germination. European Journal of Biochemistry 91(1): 263-268.

Shiomi, N. & S. Hori, 1978. Changes in proline-^{14}C metabolism in barley seedlings germinating at low temperature. Soil Science and Plant Nutrition 24(1): 121-129.

Shivakumar, A.G. & J.D. Padayatty, 1976. De Novo synthesis of 3'-ribonucleotidase in germinating rice embryos. Indian Journal of Biochemistry and Biophysics 13(3): 248-251.

Sindhu, R.K. & H.V. Desai, 1980. Regulation of agmatine iminohydrolase activity in germinating groundnut seeds. Phytochemistry 19(1): 19-21.

Sobolev, A.M., 1976. Synthesis of storage proteins in castor seeds during swelling. Doklady Botanical Sciences 229-231: 117-120.

Srivastava, S.K., V. Prakash & B.I. Naik, 1977. Regulation of diamine oxidase activity in germinating pea seeds. Phytochemistry 16(2): 185-187.

Stevens, E. & L. Stevens, 1977. Glucose-6-phosphate dehydrogenase activity under conditions of water limitation: a possible model system for enzyme reactions in unimbibed resting seeds and its

relevance to seed viability. Journal of Experimental Botany 28(103): 292-303.
Stoeck, A. & M. GRUBERT, 1978. Untersuchungen über die Endospermschleime der Samen von Cercis siliquastrum L. und Ononsis natrix L. Berichte der Deutschen Botanischen Gesellschaft 91(2/3): 369-379.
Strack, D., N. Tkotz & M. Klug, 1978. Phenylpropanoid metabolism in cotyledons of Raphanus sativus and the effect of competitive in vivo inhibition of L-phenylalanine ammonia-lyase (PAL) by hydroxylamine derivatives. Zeitschrift für Pflanzenphysiologie 89(4): 343-353.
Subramanian, V., 1972. Biochemical events during early germination of mungbean and white runner bean seeds. Thesis - Katholieke Universiteit te Leuven; Faculty of Agronomy, p. 80.
Takaiwa, F. & S. Tanifuji, 1978. Development of RNase activity in embryonic axes of germinating pea seeds. Plant and Cell Physiology 19(8): 1507-1518.
Takaiwa, F. & S. Tanifuji, 1979. RNA synthesis in embryo axes of germinating pea seeds. Plant and Cell Physiology 20(5): 875-884.
Tanaka, T. & M. Yamada, 1979. A phosphatidyl choline exchange protein isolated from germinated castor bean endosperms. Plant and Cell Physiology 20(3): 533-542.
Tárrago, J.F. & G. Nicolás, 1976. Starch degradation in the cotyledons of germinating lentils. Plant Physiology 58(5): 618-621.
Tayal, M.S., C.B. Sharma & R. Gopal, 1978. The effect of some growth regulators on phosphorus metabolism in germinating seeds of fenugreek. Indian Journal of Plant Physiology 21(1): 17-22.
Tillberg, E., 1977. Indoleacetic acid levels in Phaseolus, Zea, and Pincus during seed germination. Plant Physiology 60(2): 317-319.
Tully, R.E. & H. Beevers, 1976. Protein bodies of castor bean endosperm. Plant Physiology 58(6): 710-716.
Tully, R.E. & H. Beevers, 1978. Proteases and peptidases of castor bean endosperm. Plant Physiology 62(5): 746-750.
Uebelmann, G., 1978. Samenkeimung bei Trigonella foenum-graecum L.: Aufnahme der beim Galaktomannanabbau in Endosperm freiwerdenden Zucker durch den Embryo. International Journal of Plant Physiology 88(3): 235-253.
Umezurike, G.M. & F.A. Numfor, 1979. Changes in the content of starch and activities of some enzymes of carbohydrate metabolism in cotyledons of germinating seeds of Voandzeia subterranea. Journal of Experimental Botany 30(116): 583-588.
Venugopal, K. & G. Rama Rao, 1978. Available lysine and vitamin C in germinated blackgram (Phaseolus mungo) seeds. The Indian Journal of Nutrition and Dietetics. 15(1): 9-11.
Vick, B. & H. Beevers, 1977. Phosphatidic acid synthesis in castor bean endosperm. Plant Physiology 59(3): 459-463.
Villaneuva, V.R., R.C. Adlakha & A.M. Cantera-Soler, 1978. Changes in polyamine concentration during seed germination. Phytochemistry 17(8): 1245-1249.
Vorobjev, N.V., 1976. Metabolism of phosphorus compounds during the initial period in generation of rice in connection with different vitality of the seeds. Soviet Plant Physiology 23(4): 651-656.

Vorobjev, N.V., 1977. Intensity of respiration and activity of dehydrogenases in embryos in the processes of rice seed germination. Physiology and Biochemistry of cultivated plants 9(3): 279-284.

Vozzo, J.A., 1978. Carbohydrates, lipids and proteins in ungerminated and germinated Quercus alba embryos. Forest Science 24(4): 486-493.

Vu, C. van & R.H. Biggs, 1978. Sterols in germinating seeds and developing seedlings of longleaf pine, Pinus palustris. Physiologia Plantarum 42(3): 344-350.

Vu, C. van & R.H. Biggs, 1979. Effects of inhibitors on the biosynthesis of sterols, reducing sugars, and chlorophyll, and the development of isocitrate lyase in germinating seeds of longleaf pine, Pinus palustris Mill. Plant Science Letters 16(2/3): 255-265.

Wainwright, I.M. & I.P. Ting, 1976. Microbody malate dehydrogenase isozyme in cotyledons of Cucumis sativus L. during development. Plant Physiology 58(4): 447-452.

Wani, A.A. & S.M. Hadi, 1976. A nuclease specific for depurinated DNA in germinating pea seeds. Indian Journal of Biochemistry and Biophysics 13(3): 307-309.

Wani, A.A. & S.M. Hadi, 1979. Partial purification and properties of an endonuclease from germinating pea seeds specific for single-stranded DNA. Archives of Biochemistry and Biophysics 196(1): 138-146.

Wankhede, D.B., R. Saroja & M.R. Rao, 1977. Changes in carbohydrates and the activity of α-galactosidase, pentosanase and lipase during germination of groundnuts (Arachis hypogea). Journal of the Science of Food and Agriculture 28(2): 167-172.

Weber, E., K.H. Süss, D. Neumann & R. Manteuffel, 1979. Isolation and partial characterization of the protein body membrane from mature and germinating seeds of Vicia faba L. Biochemie und Physiologie der Pflanzen 174(2): 139-150.

Wharfe, J. & J.L. Harwood, 1979. Lipid metabolism in germinating seeds purification of ethanolamine kinase from soya bean. Biochimica et Piophysica (L) 575(1): 102-111.

Wilp, R. & R. Buschbeck, 1978. Veränderungen der Isoenzymmuster der Amylasen in Karyopsen von Winterroggen, Secale cereale, während der Reifung und Keimung. Biochemie und Physiologie der Pflanzen 173(3): 270-278.

Xavier-Filho, J. & A. Newton do Monte Negreiros, 1979. Distribution of proteinase inhibitors in Vigna sinensis seeds. International Journal of Plant Physiology 95(5): 423-430.

Yamada, M., T. Tanaka, J.C. Kader & P. Mazliak, 1978. Transfer of phospholipids from microsomes to mitochondria in germinating castor bean endosperm. Plant and Cell Physiology 19(1): 173-176.

Yamamoto, N., 1977. Protein synthesis in the embryos of Pinus thunbergii seed. I. Influences of imbibition of seeds in the dark. Plant and Cell Physiology 18(2): 287-292.

Yamamoto, N. & S. Sasaki, 1977. Protein synthesis in the embryo of Pinus thunbergii seed. II. An induction by light. Plant and Cell Physiology 18(4): 783-789.

Yamamoto, N. & S. Sasaki, 1979. Protein synthesis in the embryo of

Pinus thunbergii seed III. The Botanical Magazine Tokyo 92(1025): 31-38.
Yamasaki, Y. & Y. Suzuki, 1979. Purification and properties of an α-amylase from germinating seeds of Phaseolus vidissimus Ten. Plant and Cell Physiology 20(3): 553-562.
Yoshida, H. & G. Kajimoto, 1977a. Changes in glycolipids and phospholipid compositions in cotyledons of germinating soybeans. Agricultural and Biological Chemistry 41(10): 1857-1863.
Yoshida, H. & G. Kajimoto, 1977b. Changes in lipid component and fatty acid composition of axis and root of germinating soybeans. Agricultural and Biological Chemistry 41(12): 2431-2436.
Yoshida, H. & G. Kajimoto, 1978. Fatty acid distribution in glycolipids and phospholipids in cotyledons of germinating soybeans. Agricultural and Biological Chemistry 42(7): 1323-1330.
Yoshida, K. & K. Sasaki, 1977, Changes of template activity and proteins of chromatin during wheat germination. Plant Physiology 59(3): 497-501.
Youle, R.J. & A.H.C. Huang, 1976. Protein bodies from the endosperm of castor bean. Plant Physiology 58(6): 703-709.
Zimniak-Przybylska, Z. & J. Przybylska, 1976. Interference of Pisum seed albumins with detecting amylase activity on electropherograms: an apparent relationship between protein patterns and amylase zymograms. Genetica Polonica 17(2): 133-138.

Germination of seeds

A. Lovato

Of all the branches of science and technology on seeds, the major number of investigations are undoubtedly being carried out on seed germination. These investigations deal with many subjects: morphology, physiology and biochemistry of seed germination; factors influencing seed germination and seedling development in the laboratory and field; the study of laboratory methods to be used for tests of germination capacity. The rate of research is so great that it is extremely difficult to keep up to date with the progress. This review, with the material grouped according to subject matter, covers more than 250 papers selected mainly from those published between 1977 and 1979.

Seed morphology, microscopic and fine structure studies

The morphological studies found in the more recent literature include a wide selection of investigations ranging from the observations of Ambegaokar & Johri (1977) on the sequences of embryo genesis and subsequent development of seed *Triticale*, to a study of the embryological reasons for the low germination capacity of monocarpic tetraploid seed of sugar-beet by Jassem et al. (1978). Particular attention was devoted by Smart & O'Brien (1979) to the development of the scutellum and its histochemistry in wheat, barley, oat and ryegrass. The histological studies of Nishimura & Maeda (1977) on the origin of the callus in rice seeds revealed the physiologically important role of the plumule, interacting with mesocotyl and scutellum tissues, induction of callus. Sequences of changes in the surface appearance of lettuce endosperm, its degradation and the occurrence of cracks and pits during imbibition were observed with a scanning electron microscope and described by Pavlista & Valdovinos (1978).

Structural studies on germination in *Apium graveolens* made by Jacobsen & Pressman (1979) demonstrated that the stimulus for endosperm breakdown during germination may be a gibberellin and emanates from the embryo in response to light. Studies of Sett & Ganapathy (1977) in *Luffa cylindrica* indicated that the development of the peg, a protuberance on the concave side of the root-shoot transition zone of Cucurbitaceae, may be due to auxin-like factors from cotyledons. According to Niimi (1978), the initiation of a bulb primordium in the isolated tulip embryo is stimulated by a low temperature (4 °C) and the subsequent development is accelerated by a high temperature (24 °C).

Seed germination is accompanied by a marked increase in respiratory activity of the cotyledons or endosperm in the early stages, and by a gradual decrease in the later stages, which is related to a reduction in the function of the mitochondria. Nakayama et al. (1978) found that mitochondria isolated from 5-day-old pea cotyledons had lower respiration activity than those from 1-day-old ones and did not respond to exogenous adenosine diphosphate (ADP). Complex fine structural changes occur in seeds during germination. Lord (1978) observed that the formation of organelles as glyoxysomes and mitochondria in germinating castor bean endosperm was preceded by proliferation of the membrane system of the endoplasmic reticulum (ER). Takeba & Matsubara (1977) found that disappearance of the small fat bodies in lettuce seeds started about 4 h after imbibition at 20 °C and ended before radicle elongation began. But it was completely inhibited at 35 °C even in red light, which, however promoted their disappearance up to 27 °C.

Nayyar & Mohan Ram (1977) found that digestion of starch started from the middle of the germinating seed (cotyledons) of *Cajanus cajan* and that the hydrolysing enzymes were synthesized in ER. After succesive treatments of *Pinus halepensis* seeds with Hg^{2+} and Cl^{-} mitosis was precocious, lipolysis and proteolysis were accelerated and organelles were formed: these phenomena were particularly evident in the radicle (Pargney & Thalouarn, 1978).

Physiology and biochemistry of germinating seeds

The first step in seed germination is imbibition, during which the speed and amount of water uptake varies with several factors inside and outside the seed. Examples of internal factors are the sugar and amylopectin contents of seeds. Smooth peas take up less water but do so more quickly than wrinkled ones (Richter, 1978). On the other hand, the water uptake of the embryo tissue is inhibited by dormancy hormones, e.g. abscisic acid, which was found by Schopfer et al. (1979) primarily to inhibit water uptake by embryo tissues of *Sinapis alba* rather than to control synthesis of deoxyribonucleic acid (DNA), ribonucleic acid (RNA) or protein synthesis. Inhibitory or stimulatory substances (abscisic acid, gibberellins, fusicoccin) have a direct effect on the transmembrane electric potential of cortical cells of the embryonic axis of germinating seeds (Ballarinx-Denti & Cocucci, 1979).

Pigments in the seed may produce energy in the very early stage of germination. Edmiston (1979) found that pigments in seeds and in seedlings grown in the dark of *Sinapis alba* form a complex that absorbs ultra violet radiation and utilizes it to produce energy for growth before visible light can be utilized by the plant. According to Mukherji & Biswas (1979), the synthesis of pigments in cotyledons of intact embryos and in excised cotyledons of *Phaseolus aureus* was stimulated by penicillin, which interacted with benzyladenine in formation of chlorophyll, carotene and xanthophyll.

Although it is well known that synthesis of gibberellins de novo in the embryo is necessary for germination, this does not exclude the necessity of cytokinins and gibberellins stored in dry seeds for initiation of germination and their active involvement in the early processes. This theory is supported by the findings of Thomas et al. (1978) who observed that gibberellins, present mainly in dry seed embryos of wheat, were converted from a 'bound' form insoluble in ethyl acetate into a soluble one during seed hydration. The content of such gibberellins is then increased by some seed treatments, like vernalization (Kolbasina, 1978).

On the involvement of cytokinins in germination, Smith & Van Staden (1978) suggested that cytokinin glucosides in the endosperm of *Zea mays* are transported to the embryo where they are hydrolysed and utilized. During the first four hours of imbibition, the content of cytokinins increases because of free cytokinin synthesis, but later decreases after cells start to divide; the newly synthesized cytokinins are apparently consumed more rapidly than they are synthesized (Julin-Tegelman, 1979). Exogenous cytokinin also stimulates expansion of cotyledons and accelerates their development, but continuous application of cytokinin seems not to be prerequisite for optimum growth, as shown by Longo et al. (1978) with benzyladenine applied to water-melon seeds.

In seeds of several legumes a high proportion of the reserve material (300 g/kg in fenugreek) consists of galactomannan polysaccharides, which either regulate the water balance during germination or serve as a substrate reserve for seedling germination (Reid & Bewley, 1979). In *Acacia iteaphylla*, Ashcroft & Murray (1979) observed that the mobilization of reserve proteins (albumins and globulins) occurred slowly during the first two weeks of germination but rapidly in the third week, corresponding to the development of photosynthetic capacity in the cotyledons and the initial expansion of the primary leaf.

Other topics studied are as follows. Aldasoro & Nicolas (1980) studied the relationship of glycolytic and fermentative activities and enzymic activity of phosphoenolpyruvate carboxylase with accumulation of malate during the first hours of germination in *Cicer arietinum*. They found an increase in concentration and activity during the first 24 h of imbibition. Avadhani et al. (1978) found that rice seed germinating in darkness at low or zero concentrations of oxygen in stagnant solution showed complete inhibition of root growth, stimulation of coleoptyle elongation and high content of malate and of K^+ and a high alcoholic fermentation. By raising the O_2 concentration to 100 %, Girton (1979) increased O_2 input and CO_2 output in maize root-tip segments and in germinating grains of rice and buckwheat; but rice root-tip segments exhibited no such increase. After applications of hydroxyurea to intact seeds of *Agrostemma githago* and *Vaccaria pyramidata,* Hecker (1978) observed a strong inhibition of ^{3}H-thymidine incorporation into DNA, but the germination rate was only insignificantly reduced, whereas seedling development was strongly affected. So he concluded that DNA synthesis is not a prerequisite for radicle extension. Studying the effects of barbiturates on rice seedlings, Kordan (1978) found that the relationship between liquid/aqueous partition coefficients of barbiturates and their interference with normal mitochondrial oxidative metabolism seemed to be a factor influencing the effects of these compounds on germination and seedling behaviour. Srivastava & Kooner (1978) studied the physiological and biochemical effects of temperature on *Phaseolus aureus* seed germination in order to find the best conditions for seedling growth and to test the validity of mobilization efficiency as a criterion of seedling vigour. Hendricks & Taylorson (1979) studied the transitions at 28-32 °C in membranes of various kinds of seeds and their influence on germination and subsequent growth. Nikolaeva & Yankelevich (1979) studied the effect of low temperature and found that the activity of enzymes in yellow acacia seeds after four days of swelling at 15 °C increased with the increase in germination temperature without reaching the level of the enzymatic activity of seeds immediately placed at 25 °C.

Seed quality

The germination percentage and seedling growth vary with the quality of seed (e.g. seed size, nutrition, maturity, moisture content). In order to get good viable seeds with high potential germination and vigour, adequate time has to be allowed for the seeds to progress through and complete their maturation on the mother plant. Full maturity of seed was found to be important in *Foeniculum vulgare* (Guy, 1979a, b), *Phaseolus aureus* (Padmaja Rao et al., 1978), *Gossypium hirsutum* (Minton et al., 1979) and in *Helianthus annuus* (Zimmerman & Zimmer, 1978), where close positive correlations were found between seed maturity and seed germination. In contrast immature seed from still green pods of *Theobroma cacao* gave more vigorous seedlings than those from traditional ripe pods (Adenikinju, 1978). Seed nutrition is also important. Thomas & Raper (1979) found that an increase in nitrogen applied to the mother plant reduced the germination time and enhanced the uniformity of germination of tobacco. seed.

The influence on germination of the position of *Apium graveolens* seed on the mother plant was studied by Thomas et al. (1979) who found that the seeds produced in primary umbels were heavier than those of other umbels, but did not germinate better because about 30 % of the seeds lacked embryo or contained very small shrivelled embryos. In *Valerianella locusta*, however, seeds below 1.75 mm in diameter were without embryos or had underdeveloped embryos and germinated less than bigger ones (Kretschmer, 1978). Evidence for direct correlation between seed size and germination rate was also found by Rytko (1977) in *Dactylis glomerata* and Okada (1978) in green panic, but not by Ahring & Todd (1978) in *Cynodon dactylon*. In certain situations hard-seededness is a beneficial characteristic for seeds remaining on mature plants exposed to extended periods of warm and humid weather because their resistence to moisture reabsorption maintaines their viability better (Potts et al., 1978). However, seed viability can be markedly affected by various causes, e.g. cracks, which occur more often in large-seeded species than in small-seeded ones and in the largest seeds of a given species more than in small seeds (Pana, 1977).

Certain storage conditions may also affect viability. 'Storage fungi' of the genera *Aspergillus* and *Penicillium*, found in dried and undried stored grain of rye and oats, were considered responsible for the observed reduction of viability (Narkiewicz-Jodko, 1978). Free formaldehyde, released from seeding magazines constructed with bonded particle-board, was suspected to reduce germination and emergence of wheat seed stored in such containers (O'Brien & Orth, 1978).

The more ready the seeds for germination the less the risk that their germination and field establishment may be adversely affected by unfavourable environments and soil conditions. Starting from these considerations, many authors have recently studied ways of 'invigorating' seeds by pretreatments in order to obtain pregerminated or, better, 'primed' seeds (Heydecker, 1978), that is seeds which have completed the initial stages of germination before they are sown. Using the PEG 6000 technique, Guy (1978a) has studied the effects of pregermination on seed of 14 field and garden crops, finding a general and significant reduction of germination time. The promising results of such studies have led to a new technique of sowing pregerminated seeds with fluid drills (Currah, 1978): higher emergence rate in the field, and improved plant growth with greater marketable fruit yield of bush tomatoes were obtained by Gray et al. (1979).

Ecological conditions

In normal sowing and germination conditions and with seed having normal vigour in the field and in laboratory substrates, the orientation of the seeds (i.e. with embryo root pointing up, down or sideways) sometimes has an influence on the rate of sprout emergence, but does not have any substantial influence on total germination. The orientation of underground and aerial organs is in fact regulated by positive geotropism and heliotropism of the respective organs at the cellular level (amyloplasts) rather than by gravity. In the absence of gravity the root grows in the direction in which the seed was oriented and the shoot in the opposite direction, along the embryonix axis, as was observed for pine seeds in weightless conditions on board the 'Kosmos-782' (Platonova et al., 1977b).
The seed bed, i.e. soil compaction, nutrient status and moisture regimes exerts a marked influence on emergence and seedling growth. High soil moisture and high nutrient content reduce the adverse effects of soil compaction on seedling root and shoot development (Fawusi, 1978). The sanitary conditions of the seed bed can also have an influence on germination. In soils heavily infested with certain species of *Alternaria, Colletotricum, Helminthosporium, Fusarium,* metabolites excreted by the fungi may be partially responsible for failure of seed germination (Abraham, 1978). On the other hand, there are plants whose germinating seeds and roots exude gaseous and volatile metabolites which stimulate germination of spores of such soil fungi (Afifi & Abdulla, 1978).
The emergence and seedling growth of a species may often be regulated in other ways than by specific response to environmental factors. Takahashi (1978) studied the germination and growth behaviour of barnyardgrass grown in combination with Italian ryegrass and showed the effects of competition.
The presence of gaseous substances or any elements over the natural concentration in the atmosphere or in the soil, especially from industry, may influence seed germination and seedling growth. The effect of air pollutants on germination of *Xanthium pennsylvanicum, Amaranthus retroflexus, Brassica kaber, Ambrosia trifida* and *Glycine max* were studied by Chakrabarti (1979) who found that SO_2 and NO_2

at 24 ml/m^3 were particularly toxic and strongly inhibited germination, whereas, surprisingly, CO increased the germinablilty of all the species tested. The toxicity of Hg^+ for germinating turnip seed reduced by the presence of SO_4^{2-} and SeO_4^{2-} which are normally associated with Hg^+ both in geothermal emission and in combustion products (Spencer & Siegel, 1978). An increase in the concentration of nickel, an element found in many soils, may result from its continuous deposition on the soil surface as an air-borne pollutant. Low concentrations in air of $NiCl_2$ (up to 0,05 mol/m^3) stimulate rice seed germination, while higher concentrations (up to 100 mol/m^3) inhibit it as well as the catalase, peroxidase, indoleacetic acid (IAA) oxidase and ascorbic acid (AA) oxidase (Das et al., 1978).

Effect of insecticides, fungicides, herbicides and fertilizers

The increased use in agriculture of chemical products like pesticides, herbicides, and fertilizers raises the question whether such compounds influence seed viability and germination.

For example, Visalakshy & Nair (1978) found that phorate granular insecticides significantly decreased germination of cowpea seed and were toxic to the cotyledons. Treatment of wheat with the complex pesticide thirahexalin depressed root growth and cell numbers in the meristematic zone, while increasing cell size, nucleus volume and content of DNA per nucleus (Vintilă et al., 1979). It may happen that insecticide treatment, though not affecting seed germination, may reduce seedling development and induce cytological abberrations, as was found by Singh et al. (1979) in barley seed treated with six systemic and three non-systemic insecticides. No adverse effect of coating pea seeds with a solution of methiocarb at 3.5 g/l, a successful repellant of small birds for newly sown crops, has been observed by Burgmans (1979) in laboratory and field tests on germination. Fungicide treatment with Panogen (content in seeds 2 g/kg) stimulated germination of barley, breaking seed dormancy slightly more than gibberellic acid (Guy, 1978b). Not only insecticides but also certain substances used as carriers for fungicides may inhibit germination. Among these, some organic solvents, such as acetone, dichloromethane and ethanol, reduced the germinability of cotton, bean, pea, soya and sugar beet to varying extents depending on length of immersion, seed type and solvent (Lewis et al., 1979). Certain herbicides applied to the crop soil or present in the soil as accumulated residues can inhibit seed germination and seedling growth of some species. Raza & Saiprakash (1979) found that the germination of *Phaseolus radiatus* was strongly inhibited when treated with dichlorophenoxyacetic acid (2,4-D) at concentrations of 500 and 1000 mg/l; at 1 mg/l the hypocotyl length and radicle growth was reduced and amylase activity fell by up to 30 %. Herbicide (including Eptam, Gesagard 50, Igran, Treflan, Fasolin) affected seed germination of bean, seed yield and phosphorus, potassium, calcium and fat content in the seeds, but did not affect either the 1000-seed weight or their contents of protein and nitrogen (Guta et al., 1977). In the laboratory, the herbicides Amiben and Treflan at concentrations over 5 mg/l inhibited germination of soya seeds and caused radicle and hypocotyl malformations (Ciorlăus, 1977). By

progressively increasing prometryne concentration (20-1600 mg/l), Pahwa et al. (1979) reduced root and coleoptile extension in non-saline and saline media.
Soil fertility and fertilizers may influence seed germination and seedling growth. Results may provide practical information for agriculture. For example, chloride in fertilizer depressed germination to some degree in several crops (lettuce, cauliflower, sugar-beet, wheat, maize), so that perhaps chloride-free fertilizers should be used for such crops (Tinker et al., 1977). For nitrogen fertilizer, not only dose but also form seems important in wheat germination. Raza & Tiagi (1977) found that concentrations of nitrogen in different forms for maximum germination of wheat seed was as follows: of ammonium phosphate 100 mg/l, ammonium sulphate 200 mg/l, urea 400 mg/l, thiourea 600 mg/l, ammonium nitrate 900 mg/l, higher concentrations significantly reducing germination. Magnesium deficiency may reduce germination ability and an external supply of magnesium to seeds can stimulate germination (Kiss, 1979).

Temperature

Seed germinability, measured by the final percentage of germinated seeds, varies with incubation temperature. This temperature dependence varies with seed type. The high and low limits in temperature (above or below which the seeds do not germinate) and the optimum temperature range (within which the highest and fastest germination rate is found) are easily determined empirically. The vast scientific literature confirms that these temperature parameters vary not only with species, variety, genotype, origin and degree of maturity of the seed but also with the physiological conditions of the seeds before germination.
The effects of temperature on seed germination are complex because they involve a series of bioenergetic processes whose intensities are a function of temperature as is seen from studies on the distribution of the relative frequencies of seed germination along the isothermal incubation time. Germination, seen as a complex thermal-biological problem, has also been studied recently by Laboriau (1978), Hageseth (1978) and Nishiyama (1978) on the basis of proposed theoretical equations, and kinetic and thermodynamic models.
The temperature dependence of the germination process, including the determination of optimum and suboptimum temperatures, can be studied analytically by investigating small increments in incubation temperature over a very wide temperature range. This can be easily done on plates with a thermal gradient as described recently for tests in sand (Clegg & Eastin, 1978).
Because the germination of individual seeds is never perfectly synchronized, the evaluation of and comparison between the various temperatures studied can be made from an examination of the distribution of the relative frequencies of seed germination along the isothermal incubation time. According to Laboriau & Pacheco (1978), the best temperatures should be those that give rise to unimodal distributions of germinating frequencies, whereas suboptimal temperatures give polymodal distributions. Unimodal germination rhythms can, however, be modified to bimodal or multimodal types by inter-

fering factors unrelated to the seeds, as was found by putting *Xanthium pennsylvanicum* seeds in contact with some osmotic solutions such as mannitol (Satoh & Esashi, 1978) or by exposing them to chilling or anaerobic induction (Esashi et al., 1978).

Temperature requirements for germination differ between species and are so regular that one can observe a relation between germination temperature and the classification of species. Kawanabe (1979), studying the germination behaviour of seed samples from 27 grass species representing 25 genera, 12 tribes and 5 subfamilies, managed to distinguish between species that germinate at 6 °C within 30 d but not at 37.5 °C and those which germinate at 37.5 °C but not at 6 °C. Thompson (1979) grouped a number of bedding plants in relation to the germination temperature and the time needed to reach 50 % of maximum germination.

Some studies have been done on temperature requirements for the germination of different species. For *Sorghum bicolor* L., Aisien & Ghosh (1978) found that germination occurred between 15 and 30 °C with an optimum at 22 °C; *Crotalaria juncea, C. sericea, C. retusa* and *C. mysorensis* displayed temperature optima ranging from 20 to 30 °C (Pandey & Sinha, 1978). Different optima were observed in tree species of *Papaver* spp.: 21-26 °C for *P. bracteatum*, 18-30 °C for *P. orientale* and 13-33 °C for *P. sommiferum* (Bare et al., 1978). Studies by Daphni & Negbi (1978) on germination kinetics of different *Prosopis farcta* populations showed that the rate of germination increased with temperature from 15 to 30 °C and then decreased; in most populations, the start of germination was delayed at temperatures below 40 °C and the final germination percentage seemed unaffected in the temperature range 15 to 40 °C. For kauri (*Agathis australis*) seed, a table of temperature gradients showed that germination occurs between 10.5 and 36 °C, although adequate germination and initial growth was found only between 19.5 and 27.5 °C with an optimum at, or close to, 25 °C (Barton, 1978).

Optimum temperature for seedling development refers to the seedling in its entirety since in certain species root and shoot have different sensitivities to temperature: in apple seedlings, the optimal temperature for root growth was found to be markedly lower than that for the hypocotyl (Perino & Come, 1979).

Temperature requirements for germination also differ between cultivars: significant interactions between cultivar and temperature were observed for germination and emergence rates in *Glycine max* varieties incubated at 10 to 35 °C (Hopper et al., 1978) as well as for some soya genotypes tested at 32 and 38 °C (Emerson & Minor, 1979). Significant differences for low temperature were found among *Zea mays* cultivars by McAdam & Hayes (1978), using a thermogradient plate, which the authors found to be a simple means of finding cultivars most suitable for growing in cold regions. Low suboptimal temperatures also were used to indicate genetic differences among pepper (*Capsicum*) species (Gerson & Honma, 1978) and in sunflower seeds rich in oil (Cseresnyes, 1977).

The year of production has an influence on the germination behaviour of the various cultivars of a single species: Townsend (1977) observed that even if the optimum temperature for *Astragalus cicer* seed is 15-25 °C, several polycross progenies were able to germinate

quickly even at 5-20 °C and some at 5-15 °C, depending on the year of production.
Temperature requirements can even vary noticeably in relation to dormancy and to different storage, pretreatment or other germination conditions. Both naked and unhulled dormant seed of *Heteropogon contortus* that do not germinate for 8 to 12 months at 21 °C germinated optimally at temperatures between 30 and 35 °C (Tothill, 1977). Various changes may occur in temperature limits if the seeds are kept dry or moist at low or high temperature, exposed to light or buried. Roberts & Lockett (1977) found that constant temperatures of 20, 25 or 30 °C were ineffective for *Solanum dulcamara*, but actually promoted almost complete germination after the seed was buried 5 cm deep outdoors in winter, stored moist at 4 °C for 1 month or moistened with gibberellins. Although the optimum temperature for the germination of *Chloris* and *Eragrostis* seeds is 28 °C, Shimizu et al. (1977) substantially promoted germination at the low temperature of 18 °C after a pre-incubation at high moisture and high temperature (28 °C). Gilbertson-Ferriss & Wilkins (1977) found that optimum temperatures for a most rapid and uniform germination of *Freesia refracta* were 15.5-18.5 °C under clear polyethylene and 13-21.5 °C under black polyethylene. Studies on residual effects of high-temperature treatments (28-41 °C) on imbibed *Apium graveolens* seeds in the dark, showed that germination at 22 °C was more than for seed not treated (Biddington & Thomas, 1979). Incubations at high (35 °C) and low (10 °C) temperature for 10-24 h delayed the germination of flax seeds at 30 °C, whereas high-temperature treatment enhanced it and the low temperature delayed it in sesame (Salehuzzaman & Pasha, 1979) and in *Prosopis cineraria* (Mohnot & Chatterji, 1978)
Furthermore there are seeds for which the optimum constant temperature decreases during storage, probably because after-ripening progressively reduces the dormancy, as for cassava (*Manihot esculenta*) (Ellis & Roberts, 1979), who also found that temperature alternation itself can be stimulatory because even if the range of the alternation does not include the optimum constant temperature value, the percentage germination can be higher than at any constant temperature within the range. On the other hand, the range of temperatures that allowed the germination of cold stored seeds of *Quercus robur* shifted towards higher temperatures with increasing storage time (Levert & Lamont, 1979). Gomes & Kretschmer (1978) compared three high temperatures (30, 38 and 46 °C) alternated daily with 25 °C for 8 h for scarified and unscarified seed of tropical legumes (*Macroptilium, Aeschynomene, Centrosema, Stylosanthes, Desmodium, Calopogonium*). They noticed that high temperature could function as scarification agents for some species. The same happened to untreated *Ischaemum magnum* seeds, incapable of germination at 25-30 °C in the field or laboratory; with a warmer regime (25-35 °C), seed germinated in a cabinet as well as when seeds were treated with boron and gibberellic acid (Ng & Wong, 1977).
Much of the literature reports the favourable effect of daily temperature fluctations on seed germination: many kinds of seed germinate better with alternating temperatures than with constant ones. Seeds of *Xanthium pennsylvanicum* would not germinate

at constant temperature below 25 °C, but germinated to maximum extend with alternating temperatures of 16 and 23 °C (Esashi & Tsukada, 1978). Different temperature of incubation can have less influence when the temperatures tested, both alternated or constant, fall within the optimum temperature range for the species, as was observed for *Oryza sativa* (Singh et al., 1973) and *Lathyrus sylvestris* (Chirco et al., 1978). Seeds of *Luffa cylindrica*, a tropical species that germinated poorly at constant temperatures of 15 and 31 °C, fairly at 21 °C but not at 41 °C, gave high germination rates with those combinations of daily alternating temperatures (15-41 and 21-31 °C) (Okusanya, 1978).

Moisture, water stress and salinity

The availability of sufficient water is essential for the imbibition of seeds, their germination and seedling growth. The rate of seed swelling, germination, root and shoot extension depends on the water potential of the soil. Sequential studies of such succesive biological events have been made by non-destructive serial neutron radiography (Willatt & Struss, 1979).
Insufficient moisture in the seed bed is probably the prime cause of the failure of seeds to germinate. The practical consequence of a dry seed bed for rate of imbibition and germination of seeds has been studied and discussed by Collis-George & Melville (1978) with a model simulating conditions where water was transported only as vapour.
The amount of moisture required for germination varies considerably with species, variety, seed size and factors that influence moisture content of the seed bed. For example, reductions in matrix potentials of soil (-0.2 to -1.3 MPa) were found to decrease the proportion of germination and the rate of emergence of *Bromus tectorum* seeds, especially with low soil temperature (5-15 °C) (Thill et al., 1979a). Varietal differences in water requirements for germination were found in *Gossypium hirsutum* (Sheikh & Mall, 1978).
Small seeds, which can not be planted very deeply, are particularly affected by loss of moisture from the surface layers of the soil. For the same moisture content of soil, the maximum depth of planting depends on the type of soil (Padilla et al., 1978) and on the tolerance of the germinated seed to lack of moisture, which must be known in order to choose species and varieties adapted for establishment on soils whose surface layers are rapidly drying. Investigations on the effect of temporary water stress on pregerminated seeds have been made for *Dichanthium sericeum* (Watt, 1978), *Agropyron desertorum* and *Elymus junceus* (Hassanyar & Wilson, 1979), showing that partially germinated seeds (at a very early stage) can maintain their capacity to develop seedlings after temporary desiccation.
The influence of water stress on the physiological and biochemical processes in germinating seeds (seed swelling, rate of respiration, starch accumulation), the rate of germination, seedling growth and rate of emergence have been studied in the laboratory with osmotic solutions like dextran, sodium chloride, sodium sulphate, mannitol, polyethylene glycol with different osmotic potentials (Sheoran et al., 1979; Turn et al., 1979). The results from such studies are

often compared to the results from soil media in order to determine the solutions' capacity to simulate 'true' drought conditions (Till et al., 1979b).

Differences in varietal response to moisture stress have been found in some species. This is undoubtedly of considerable agronomic interest in selection of drought-tolerant and salt-tolerant cultivars. Ashraf & Abu-Shakra (1978) found that the total germination of wheat cultivars was not affected by osmotic tension up to 1.2 MPa, but was significantly reduced at 1.5-1.8 MPa. However speed of germination, root growth and respiration rates were inversely related to moisture stress and for the parameters studied, significant differences were observed among cultivars. In *Sorghum bicolor*, Mali et al. (1978, 1979) found that there was not only a specific critical water potential for soil (300-620 kPa) and critical seed hydration (mass fraction of water in seed 210.7-344.2 g/kg) for germination of 14 cultivars, but also that the total amount of water absorbed during germination at different moisture potentials differed considerably for seeds of the same varieties. Also for *Helianthus annuus* and *Carthamus tinctorius* germination at different osmotic potentials with NaCl (up to 2.51 MPa) and polyethylene glycol 6 000 (up to 1,68 MPa) varied markedly with cultivar (Maftoun & Sapaskhah, 1978).

Regardless of the sensitivity of the seed to water stress, this sensitivity for seed germination is higher in many species than for radicle growth. Radicle growth was inhibited at lower water potentials than seed germination, with value differences of about 0.6-0.8 MPa, in turnip, calabrese, radish, parsnip, beetroot, lettuce and red clover, though not significantly in carrot and cucumber (Hegarty & Ross, 1978; Ross & Hegarty, 1979). Moreover this different sensitivity could not be usefully removed with growth regulators, such as an ethrel and kinetin combination; in fact these regulators can inhibit growth after germination (Hegarty & Ross, 1979).

Water deficits may be due to the presence of soluble salts in soil water. The consequences of soil salinity on germination of ten fodder plants and seven wheat cultivars, and on dry matter yield and chemical composition of *Hedysarum coronarium* have been recently investigated by Myers & Carloni (1978). Different concentration of NaCl and $CaCl_2$ were used to study the influence of salinity on the seed germination and seedling growth of annual medics; a progressive and consistent depression in these parameters was observed with increasing salt concentrations (NaCl more than $CaCl_2$). The depression was most marked for *Medicago orbicularis* with NaCl and *M. polymorpha* var. *brevispina* for $CaCl_2$ (Rizk et al., 1978). Plant species characteristic of Indian arid zones with different osmotic potentials of soils (*Trianthema triquetra, Cressa cretica, Suaeda fruticosa* and *Halowylon recurvum*) decreased in germination with decrease in osmotic potential (Rajpurohit & Sen, 1977). Differences in sensitivity of wheat cultivars to salinity as indicated by germination, shoot growth and α-amylase activity (which decreased with increase in salinity) were found by Ansari et al. (1977).

In salined soil (salt content 1-8 mg/kg), grain yield decreased with salt content whereas at all contents yield was greater than of the control when seeds were soaked for 24 h in salt solution with NaCl, NA_2SO_4 and $CaCl_2$ up to 30 mg/l (Mehta et al., 1979).

Light

The influence of light on germination has been recognized for a long time. There are species that germinate in the light only, others only in the dark and still others that germinate in either light or dark. Light sensitive species can be induced to germinate by the intensity and quality of the light penetrating the soil. Spectrometric measurements showed that less than 1 % of the incident radiation in the visible spectrum (wavelength between 350 and 780 nm) penetrated 2.2 mm of a soil with ped up to 1 mm in size (Woolley & Stoller, 1978). Thus an exposure to light equivalent to about one sunny day induced germination of some of the lettuce seeds which were 2 mm below the soil surface, but germination was not induced in lettuce seeds 6 mm deep. Germination of lettuce seed could also be induced by a millisecond flash of red light (Haupt & Scheuerlein, 1977). On the other hand, seeds of *Cercopia obtusifolia*, a tropical forest species that germinates only in light, were inhibited by light of the quality inside forest (Vazquez-Yanes, 1979).
As known, the sensitivity of seeds to light is regulated by a chromoprotein known as phytochrome. The functions and properties of phytochrome have been studied by Pratt (1978). Duke (1978a) has discussed the influence of this compound on changes in germination response. Studies on the different responses to light of phytochrome in different seeds are continuing, and confirm that the sensitivity of seeds during germination varies with species, variety, light quality and intermittence, temperature, imbibition and growth substances.
Genetic differences in sensitivity to promotive light at a variety level led to an investigation on the genetic behaviour of four varieties of tobacco, with the hypothesis that a double-recessive homozygous genotype was responsible for the dark germination (Watanabe et al., 1979). Variation in genetic component of germination response to red and far-red irradiation has also been suggested for some populations of wild and weed sunflowers (*Helianthus exilis, H. boulanderi* and *H. exilis* × *H. boulanderi*) (Olivieri & Jain, 1978).
Different responses to darkness, light and different temperature conditions (15 and 30 °C) were observed between different provenances (Japanese, Indian and African) of *Eleusine coracana* (Shimizu & Mochizuki, 1978).
Some seeds such as those of dwarf water-melon (*Citrullus lanatus*) germinate best in darkness with or without previous prolonged incubation in far-red light. But continuous low intensity far-red or red irradiation prevented or decreased germination (Loy & Evensen, 1979).The effect of different periods of incubation seemed to indicate that the germination of such seeds was regulated by a phytochrome between about 6-24 h of incubation at 29-30 °C.
A reversible phytochrome reaction with continuous or intermittent red and far-red irradiation and with temperature has been found in *Papaver bracteatum* seeds by Bare et al. (1978). In *Cucumis anguria*, red light and darkness promoted the seed germination and reversed the inhibition of far-red irradiation. Also white and blue light inhibited germination (Noronha et al., 1978). The same was found in *Aristida murina* where cool fluorescent white light and both red and far-red, individually or combined in certain sequences, inhibited

jermination (Ginzo, 1978).
ous short exposure of *Stellaria media* seeds to green light gave germination in darkness (Baskin & Baskin, 1979). Blue, green, ow, red and far-red inhibited germination of freshly collected non-dormant *Urgenia indica* seeds, which germinated better in ight with a photoperiod of 12-16 h (Khare, 1978). Different phyto-ome sensitivities to red, white, blue and far-red light were also observed in *Arabidopsis thaliana* seeds (Ritzel, 1977).
The existing relation between light and temperature requirements for germination in several seed species was demonstrated again by other authors. Porter & Clark (1979) found that temperature fluctuation (15-25 °C) and fluorescent light enhanced germination in both *Solanum aviculare* and *S. laciniatum*. But germination was significantly better when light and temperature were both fluctuated. In celery (*Apium graveolens*) seeds, Pressman et al. (1977) found that daily alternating temperatures would also replace the light requirement of seeds for germination. Takeba & Matsubara (1979) studied the growth potential of the embryo in New York lettuce and suggested that it is probably controlled by some thermolabile process in addition to its phytochrome(Pfr)-dependent reaction. Other factors, such as hormones and imbibition of the seeds, may interact with light. Duke (1978b) observed that the reduction of water content of the seed of *Rumex crispus* with mannitol or polyethylene glycol 6000, decreased the effectiveness of the far-red absorbing form of phytochrome in causing germination. Hormones are believed to be involved in phytochrome-mediated growth in embryonic axes: gibberellic acid stimulated elongation of axes from lettuce seeds treated with far-red but not those treated with red light, while kinetin and abscisic acid reduced axis elongation and fusicoccin stimulated it with both red and far-red treatment (Carpita et al., 1979).

Stimulatory and inhibitory substances

Besides factors like temperature, light, water and oxygen, germination is controlled by inhibitory or stimulatory chemical substances inside or outside the seed.
Effects of growth hormones and certain chemical substances on germination and seedling growth have been known for quite some time and are still the object of numerous investigations. The slow and erratic germination of several species is a serious obstacle to plant establishment in the field. So laboratory studies were prerequisite to study of germination in the soil.
Certain substances present in seeds can specifically inhibit or stimulate germination of seeds in which they occur as well as on seeds of other species. The inhibition of seed germination and seedling growth by the seed pod of *Onobrychis viciifolia* was found to be closely related to the presence in the pod of several water-soluble inhibitors (Smith, 1979). The seed coat of dormant seeds of *Bupleurum falcatum* contained two groups of substances significantly inhibiting germination of lettuce seed (Momonoki et al., 1978).
Rice hulls contain nicotinamide, which promotes rooting and growth of rice seedlings (Sanai et al., 1977). The fruit wall of *Malva verticillata* contains a water-soluble germination-inhibiting factor,

which was inactivated by heating the seed to 70 °C (Ruge et al., 1977). Seeds of certain species release substances whose effects vary with cultivar. Abdul-Baki & Stoner (1978) found that an extract from seeds of a tomato cultivar, whose seeds germinate well at low temperature, promoted germination of seed of the same and other tomato cultivars; when the extract was taken from seeds of a cultivar that germinated poorly at low temperatures germination was inhibited. Also, naturally occurring substances in plants can influence germination, shoot and root growth. Mangano (1977) showed that aqueous extracts from root and culm tissues of sorghum and maize inhibited germination and seedling growth of wheat, maize, lucerne and Italian ryegrass. Leaves and stems of lawn pennywort (*Hydrocotyle sibthorpioides*) buried in the soil or its water extract markedly inhibited germination and plant growth of turnip, tomato and *Paspalum notatum*, but not of rice (Tsuzuki & Watanabe, 1978). Growth of lettuce seedling was strongly inhibited by eluate from *Glycine wightii* seedlings (Wilson & Bell, 1978). The leaf extract of *Pithecel lobium dulce* (= *Mimosa dulcis*) stimulated germination and shoot growth of wheat (Chaurasia & Shaha, 1978). Exudate of blue-green algae enhanced germination of submerged rice seeds and growth of seedlings and, by virtue of the oxygen produced, reduced losses caused by the sulphate-reducing activity of bacteria (Jacq & Roger, 1977). Bhiravamurty et al. (1979) found that compounds of plant origin, such as the furofuranoid lignans, having piperonyl residues (fargesin, sesamin, eudesmin) retarded germination of peanut, cucumber and rice. Crop residues often contain phytotoxins, which are released or leached as part of the decomposition process and remain in the soil even if in very small concentrations. Among these, vanillic and *p*-hydrobenzoic acids at concentrations of 0.5-2.5 mmol/l inhibited germination of radish and sorghum and growth of sorghum seedlings (Einhellig & Rasmussen, 1978).

The right concentrations of hormones and growth regulator-like substances are essential in order to avoid effects opposite to those expected. Sahai et al. (1977) found that gibberellic acid (GA_3) inhibited germination of *Lathyrus aphaca* at pre-soaking concentrations below 200 mg/l, but enhanced it at 300 to 500 mg/l. Germination was promoted by up to 50 mg/l of Coumarin, Phosphon and sodium chloride, 100 mg/l of maleic hydrazide, 200 mg/l of daminozide and copper sulphate, but retarded at higher concentrations. Varshney & Baijal (1978) showed that lower concentrations (up to 20 mg/l) of GA_3 and indoleacetic acid (IAA) had favourable effects on both germination and seedling growth of *Pennisetum pedicellatum*, whereas higher concentrations (50 and 100 mg/l) were antagonistic. Morphactin concentrations of 1-100 mg/l also showed this antagonistic behaviour. Germination and seedling growth of lettuce were increased by concentrations of Vitamin K at 50-100 mg/l and seedling development only was increased by IAA of 5 mg/l; higher concentrations of either substances and their combinations inhibited germination and seedling growth (Mullick & Chatterji, 1978).

Stimulatory effects on *Prunus persica* germination were obtained with GA_3 (up to 200 mg/l), and thiourea (up to 7500 mg/l) by Hundal & Khajuria (1979), and with GA_3, 4-(indole 3-yl) butyric acid (IBA), IAA, naphthylacetic acid (NAA), 2,4-D by Damiano et al. (1976). Seed germination and seedling growth of *Raphanus sativus* were favoured by GA_3 at 25-200 mg/l inhibited by coumarin at 1.5-6.0 mg/l (Kumar & Alka, 1978). Also, seed germination of *Solanum laciniatum* and *S. aviculare* was favoured by seed treatments with GA_3 (Porter & Gilmore, 1976) and nearly 100 % germination of *S. viarum* and *S. khasianum* was obtained after pre-soaking seeds for 12-18 h with a solution of GA of 1000 mg/l + ten aminoacids each at a concentration of 0.5 mg/l (Pingle & Dnyansagar, 1979).

In *Carica papaya* 10 days after sowing, germination was significantly higher for seeds treated with GA but after 20 days this difference was not evident, whereas Ethrel and daminozide showed an inhibitory effect. None of these hormones affected seedling growth (Salazar & Perez, 1976). Both hardening before sowing and foliar applications to plants of *Atriplex hortensis* 20 days old of GA, Ethrel, chlorcoline chloride (CCC), ascorbic acid (AA) and kinetin at concentrations of up to 500 mg/l each, had considerable influence in opposite ways, on the shoot length and dry weight, depending on the hormone and concentration used (Saraswathamma, 1979).

In certain legume species treatments of the seed with hormone (GA, IAA, NAA and (phenoxyacetic acid) PAA) had no effect on germination; the seedling growth was indeed influenced, although differently depending on species and concentration, as found by Sharma & Lavania (1978) in *Vicia sativa* and *V. hirsuta* and by Ratajczak et al. (1979a, b) in lupine, broad bean and pea.

The effects of treatments with growth regulators before sowing not only on seed germination and seedling growth but also on subsequent plant development are well documented. Verma & Sing (1978) found that pre-soaking barley seed with IAA, IBA (10-50 mg/l) and GA (10-100 mg/l) increased germination, plant height, tillers per plant, grains per ear and grain yield over a control. Asthana & Srivastava (1978) found that ascorbic acid stimulated maize seed germination, whereas salicylic acid either alone or in combination with ascorbic acid inhibited germination as well as root and shoot growth. However both these acids, alone or in combination, accelerated flowering and ear emergence. On the other hand, flurenol (9-hydroxyflorine-9-carboxylic acid) compounds inhibited germination of *Trigonella foenum-graecum* (Vatsa & Sharma, 1978) but not of *Cicer arietinum* (Mohan Ram & Mazumdar, 1977), delayed seedling growth and decreased seedling weight, inducing a certain negative geotropism of roots in *C. arietinum*, and causing a general reduction in plant height, pods per plant and seed yield in *T. foenum-graecum*.

The action of stimulatory or inhibitory substances, however, is not independent of other factors, such as temperature, light and imbibition. Existing interactions between gibberellic acid and light on seed germination were clearly demonstrated for *Cactus* species by Brencher et al. (1978). Evans & Fratianne (1977) found that in light sensitive *Lepidium virginicum* seeds, gibberellin could replace red light, but the effect of both could be nullified by abscisic acid. Kinetin alone only slightly promoted germination of seeds incubated in the dark; the same happened with kinetin in combination with sub-optimal concentrations of gibberellin, whereas optimum concentrations with kinetin could reverse the inhibitory effect of certain concentrations of abscisic acid (ABA). An inhibitory action of 2,4-

dinitrophenol (DNP) and ultraviolet radiation on germination and seedling growth of sunflower in the dark was prevented in white light (Purohit, 1978). Lewak & Khan (1977) found that the early process of lettuce germination initiated by gibberellic acid or red light could be arrested by incubating treated seeds at 35 °C. Phaneendranath & Funk (1978) observed that the germination of Kentucky bluegrass (*Poa pratensis*) was almost completely inhibited above optimum temperatures, though gibberellin by itself or in combination with kinetin or 2-chloroethyl phosphoric acid (ethephon) improved germination; these effects were not found at suboptimal temperatures. In a field trial on wheat, Chhipa & Lal (1978) found that soaking seed with IAA or IBA at 200 mg/l and $NaSO_4$ at 30 mg/l tended to mitigate the adverse effect of higher salinity and sodium adsorption ratio of water, increasing germination, tillering, plant height, grain and straw yield and contents of nitrogen, phosphorus and potassium in grain and potassium in straw.

Physical and chemical pretreatments

Freshly harvested seeds of many plant species and sometimes even seeds stored for a period of time do not germinate, even under favourable conditions, i.e. with an adequate supply of water and oxygen, and suitable temperature and light. Nevertheless, such seed may be considered to be viable but 'dormant' since germination can be induced by various treatments. Dormancy, germination and longevity has recently been interpreted as physiological phases for germinating cereal seeds by Takahashi (1979). Besides physiological factors, morphological and ecological factors too are involved in determining the quiescence of seed. So many physical or chemical methods are under study for breaking dormancy and clarifying its causative factors.
Pretreatments of *Pennisetum typhoides* seeds at either low temperatures (-10 or 0 °C) or high temperatures (+40 or +60 °C) for not more than one day significantly increased germination (Raza, 1977). However, germination of *Carica papaya* seed, which is enclosed in a gelatinous aril containing inhibitors was stimulated by low temperatures (10-15 °C) only if the aril was removed; high temperatures had no effect. Air drying at room temperature for 20 d promoted germination of seed with an aril (Yahiro, 1979). After removing a portion of the seed coat above the embryo of *Zizania aquatica*, Campiranon & Koukkari (1977) obtained prompt anaerobic germination, quick seedling development and high activity of alcohol dehydrogenase. Mechanical scarification by rubbing or shaking the seed for 1 to 20 min and treatment with sulphuric acid for 1 to 15 min allowed the entry of water into seed of *Vicia sativa* and *V. hirsuta* and allowed almost complete germination (Sharma & Lavania, 1979). Studying the effect of seed soaking, Rao et al. (1978) showed that such a treatment of cotton seed advanced emergence and a higher total germination than the control. According to Perl (1979) treatment of cotton seeds at 20 °C and 95 % relative humidity for up to 12 days increased rate of seedling development: such a treatment is seen as an 'invigorating' method. Some doubts were, however, expressed by Fowler (1979) who did not find reason to recommend

soaking as a practical means of improving establishment of cotton seed. Soaking enhances germination of beet because it removes inhibitors and allows a better penetration of water and aeration of the embryo through the cells of the operculum periphery and basal pores (Coumans, 1978). Removal of the operculum or placing dormant sugarbeet with the operculum downward on moistened filter paper increased germination (Coumans et al., 1979). In *Atriplex repanda* too, soaking removed water-soluble substances from the pericarp; in this way Johnston & Fernandez (1979) improved germination.
To reduce the proportion of hard seed in legumes, electrical treatments have been tried. Nelson et al. (1977) increased germination of lucerne by exposing seeds to either electric fields of radiofrequency or to gas-plasma radiation in an electric glow discharge. Both types of treatment were more effective when moisture content of seed was low (27 g/kg) (2.7 %) than at normal moisture content (78 g/kg) and the seed quality remained as good as that of the untreated seeds. Also, microwave radiation at 2400 MHz of *Acacia* spp. seeds made the seed coat permeable to water and maximized germination (Tran, 1979).
Magnetism can affect a wide range of physiological processes in seeds, seedlings and plants. Magnetic treatments (0,3 T) increased the moisture absorption and growth rate of dry soya beans (Kawi, 1977), but had no effect on germination and seedling growth of dry wheat, barley and oats. Growth rate of barley soaked for 10 h before magnetic treatment was decreased (Gusta et al., 1978). Exposure of barley and hard winter wheat to uneven magnetic field of about 120 kA/m for 10 d at 23 °C before soaking reduced the activity of amylolytic enzyme in seed parts, whole seeds or very young seedlings (Pittman et al., 1979). The seed orientation in a magnetic field seems to influence the effect of the treatments. Cerdonio et al. (1979) found that growth rate of pea from seeds exposed for 10 d to a weak magnetic field of about 8 kA/m, was inhibited if seeds were oriented 'antiparallel', i.e. with the radicle facing the south pole of the magnet, but enhanced if parallel.
Effects of radiation below microwave frequency (2400 MHz) have been explored as mean of weed control. Seed of *Echinochloa crus-galli*, *Digitaria sanguinalis* and *Portulaca oleracea* exposed to high dosages of such radiation by Hooper et al. (1978) showed membrane breakage and disorganized organelles. Adenosetriphosphatase (ATPase) activity of tissues was absent even after a minimum dosage, although the cells appeared normal. The loss of viability can be explained by extensive disruption of cytoplasmic constituents and enzyme inactivation. However ultrasonic waves of frequency 800 kHz, and volumic power 3 and 7 W/cm^3 for 5 to 45 min stimulated germination of maize seeds, increasing mass of dry matter in 4-leaf to 6-leaf plants, and grain yield (Nagy et al., 1978).
Any kind of unnatural vibration may have differing effects on living objects such as seeds. Dadykin & Kovadlo (1979), by treating seeds of trees (*Pinus, Picea, Larix, Quercus*) and pea with vibration of 30-70 Hz at an acceleration of 4.9-14.7 m/s^2 obtained an increase in proportion and vigour of germination as well as in activity of catalase, amylase, peroxidase and succinate dehydrogenase, whereas higher frequencies of 10-15 MHz at an acceleration of 49-147 m/s^2 depressed germination and activities of amylase and

catalase. Seed of cereals (wheat, barley and rye) were sensitive to a shock-wave pressure (0,1-2 GPa) from an explosion in water: a small proportion of seed remained viable; with higher pressures, germination decreased sharply (Panov et al., 1977).
Would space flight factors have any effect on seed attributes and performance? Experiments by Platonova et al. (1977) revealed a slight but significant increase in the frequency of chromosome reorganization and in germination of diploid and autotetraploid buckwheat (*Fagopyrum esculentum*) seed after 72 d on the orbital station 'Salyut 1'. This effect could have been due to temperature increases during ascent or descent rather than to the space environments, e.g. cosmic irradiation. The radiation reaching the interior of space vehicles ranges from about a microgray to a milligray (1 Gy = 1 J/kg) and does not reach the order of gray necessary to stimulate seeds.

Specific germination and propagation methods for some species

As has already been seen from the work reported in the preceeding sections, intense research has been in progress on the influence of various factors on the germination processes of seeds and the development of the sprouts; no less intense is the work related to the study of methods of germination for individual species. These studies are particularly important not only from a scientific point of view but also from a practical standpoint because of their contribution to a better understanding of specific problems related to germination and propagation, especially of plant species whose seeds have difficult germination. The extensive and heterogeneous literature in this field makes it difficult to present a detailed examination of all this research, thus brief references only are given here.
Better and faster germination of pineapple (*Ananas comosus*) seeds was obtained by exposing them to intermittent mist for 20 s at intervals of 3.5 min than by using the conventional practice of seed scarification (Iyer et al., 1978).
Spikelet clusters of *Anthephora pubescens* were found to germinate best between 20 and 30 °C (Burger et al., 1979).
Banksia spp. germinate readily without any previous treatment, in sand at temperatures of 18-23 °C for *B. integrifolia*, 20-28 °C for *B. serrata* and 28-32 °C for *B. aemula* (Heslehurst, 1979).
The different types of beet (*Beta vulgaris*) seed presently available (polygerm, monogerm, diploid, polyploid, pelleted) have different requirements for germination in the laboratory (Lovato & Amaducci, 1970).
The possibility of regenerating red pepper (*Capsicum annuum* and *C. frutescens*) plants from excised hypocotyl and cotyledons with auxins, cytokinins and growth regulators was described by Gunay & Rao (1978).
Complete germination of pecan nuts (*Carya* spp.) was obtained in two weeks at 30-35 °C, instead of stratification of the nuts (Caminada, 1979).
Some biological characteristics of *Citrus clementine* seed and its germination behaviour in relation to the number of seeds per segment was discussed by Russo (1977a, b).

Germination of *Dystaenia takesmania* was studied by Park et al. (1977).

For propagation of geraniums by seeds, Mieseler (1979) recommends sowing the seed in a soilless mixture kept at 21-24 °C: about 80 to 90 % germination should occur in the first week, the remainder the second week.

Temperature (20-25 °C) and light requirements for germination of *Gloriosa* seed have been studied by Carow (1979).

The method of germinating *Hieracium aurtantiacum* recommended by Lincoln (1978) is as follows: seeds on top of filter paper, constant temperature of 20 °C, light, first count after 6 d and final count after 21 d.

Almost complete germination of *Humulus lupulus* was obtained with either decorticated seeds or Completely soaked seeds kept at 1-3 °C for 150 d after harvest (Suciu et al., 1978).

High germination rates of different species of *Iris* spp. were obtained after cold storage of 'chitted' seeds kept at 2 °C for 14-16 w and then incubated at 17 °C (Bartlett, 1979), or after soaking, peeling or cutting the seeds (Tamberg, 1979).

Seeds of *Ischaemum magnum* germinated well after dehulling or treatment with concentrated H_2SO_4 or KNO_3 at 1 g/l as well as germinated in a cabinet at 25-32 °C (Ng & Wong, 1977).

The maximum germination rate of seeds of *Lemaireocereus griseus* (*Cactaceae*) was in light at 25 °C, after swelling in water (Clausnitzer, 1979).

The rhythm of absorption and the influence of soil moisture, light and temperature on the germination of *Lupinus albus* seeds was accurately examined by Muntean (1978).

Germination of *Myrica pennsylvanica* is increased by GA_3 at concentrations of 500 and 900 mg/l, which reduces the stratification time (Hamilton & Carpenter, 1977).

Mathiasen (1978) studied seed germination of three destructive forest mistletoes of the southwestern United States (*Arceuthobium apachecum, A. blumeri, A. microcarpum*) and techniques of infesting *Pinus* and *Picea* seedlings with pregerminated mistletoe seeds.

The propagation by seeds of some orchids is often difficult, germination rate depending upon species and on requirements for an exogenous source of suitable carbohydrate (Harrison & Arditti, 1978; Courtis, 1978). Different factors and techniques affecting germination of some orchid seeds have been studied by Ichihashi (1978) for *Bletilla striata*, Hasegawa et al. (1978) for *Calanthe* spp., Muick (1978) for *Cypripedium* spp., Houck (1979) for *Phalaenopsis hybrids*, Arekal & Ananda Karanth (1978) for *Zeuxine strateumatica*.

Requirements for good germination of freshly harvested *Panicum clandestinum* in the laboratory were studied by Chirco et al. (1979), who obtained the highest germination rate after chilling at 5-10 °C for 7 d and with alternating temperatures of 10 and 30 °C for 6 w.

Many palm seeds germinate slowly and poorly. Among techniques suggested to improve and hasten germination are the removal of seed from fruits (Manokaran, 1979) and soaking in water plus hormone (Nagao & Sakai, 1979).

Studies on the germination of 25 *Petunia* lines showed the existence of several and sometimes conflicting dormancy states regulated by

at least two sets of genes (Jassey et al., 1979).
The germination of *Phalaris arundinacea* at 24-27 °C was significantly stimulated by chilling in light (red light during the first 3 d of soaking), ethylene, increased oxygen tension and soaking for 4 d (Landgraff & Junttila, 1979).
Freshly collected seeds of *Piptadenia falcata* (Mimosoideae) show a great imbibition capacity and germinate easily, but lose their germinating power after a month of storage (Piccolo & Thomazini, 1977).
Maximum germination of *Robinia pseudoacacia* whose seeds have a hard coat, was obtained after 9 min of scarification with concentrated H_2SO_4 (Pathak et al., 1978). Acid scarification is also essential for germination of *Rosa dumetorum* (Devoy, 1979).
Best conditions for the germination of *Silphium perfoliatum* were an incubation temperature between 22 and 25 °C for 21 d and a filter-paper substrate, or 28 d in sand (Kawahara et al., 1977).
Both 'old' and 'fresh' seed of *Talinum triangulare*, a much-cherished tropical vegetable crop of the family Portulacaceae but difficult to propagate from seeds, germinated better scarified or treated with activated carbon and thiourea (50 mg/l) (Fawusi, 1979).
Seed size seems to govern germination of *Vaccinium ashei* (Austin & Cundiff, 1978) and *V. angustifolium* (Aalders & Hall, 1979).
Stratification of *Vitis vinifera* kept in a mixture of orthocide solution at 3 mg/l and sand and peat (1:1) in polyethylene bags at 0-2 °C up to the end of January and then on filter at 27 °C gave high germination rates (Misic et al., 1978).
Studies on *Welwitschia mirabilis* seed germination and seedling establishment in natural climatic conditions were made by Sharma (1978).

References

Aalders, L.E & I.V. Hall, 1979. Germination of lowbush blueberry seeds as affected by sizing, planting cover, storage, and pelleting. Canadian Journal of Plant Science 59(2): 527-530.
Abdul-Baki, A.A. & A. Stoner, 1978. Germination promotor and inhibitor in leachates from tomato seeds. Journal of the American Society for Horticultural Science 103(5): 684-686.
Abraham, T.A., 1978. Inhibitory effect of culture filtrates of fungi on seed germination. Indian Phytopathology 31(3): 402-403.
Adenikinju, S.A., 1978. Effects of bean maturity on seedling vigour in cacao (Theobroma cacao). Turrialba 28(3): 225-231.
Afifi, A.F. & M.El-S. Abdulla, 1978. The role of gaseous and volatile root metabolites of Lupinus termis on the spore germination of rhizospheric mycoflora. Indian Phytopatholy 31(3): 334-338.
Ahring, R.M. & G.W. Todd, 1978. Seed size and germination of hulled and unhulled bermudagrass seeds. Agronomy Journal 70(4): 667-670.
Aisien, A.O. & B.P. Ghosh, 1978. Preliminary studies of the germination behaviour of guinea corn (Sorghum vulgare). Journal of the Science of Food and Agriculture 29(10): 851-852.
Aldasoro, J. & G. Nicolás, 1980. Fermentative products and dark CO_2 fixation during germination of seeds of Cicer arietinum. Phytochemistry 19(1): 3-5.
Ambegaokar, K.B. & B.M. Johri, 1977. Seed development in triticale-1.

Phytomorphology 27(2): 190-197.
Ansari, R., S.M. Naqvi & A.R. Azmi, 1977. Effect of salinity on germination seedling growth and α-amylase activity in wheat. Pakistan Journal of Botany 9(2): 163-166.
Arekal, G.D. & K. Ananda Karanth, 1978. In vitro seed germination of Zeuxine strateumatica Schlr. (Z. sulcata Lindley), Orchidaceae. Current Science 47(15): 552-553.
Ashcroft, W.J. & D.R. Murray, 1979. The dual functions of the cotyledons of Acacia iteaphylla F. Muell. (Mimosoideae). Australian Journal of Botany 27(4): 343-352.
Ashraf, C.M. & S. Abu-Shakra, 1978. Wheat seed germination under low temperature and moisture stress. Agronomy Journal 70(1): 135-139.
Asthana, J.S. & H.S. Srivastava, 1978. Effects of presowing treatment of maize seeds with ascorbic acid and salicylic acid on seed germination, seedling growth and nitrate assimilation in the seedling. Indian Journal of Plant Physiology 21(2): 150-155.
Austin, M.E. & J.S. Cundiff, 1978. Factors affecting rabbiteye blueberry seed germination. Journal of the American Society for Horticultural Science 103(4): 530-533.
Avadhani, P.N., H. Greenway, R. Lefroy & Lynda Prior, 1978. Alcoholic fermentation and malate metabolism in rice germinating at low oxygen concentrations. Australian Journal of Plant Physiology 5(1): 15-25.
Ballarin-Denti, A. & M. Cocucci, 1979. Effects of abscisic acid, gibberellic acid and fusicoccin on the transmembrane potential during the early phases of germination in radish (Raphanus sativus L.) seeds. Planta 146(1): 19-23.
Bare, C.E., V.K. Thoole & W.A. Gentner, 1978. Temperature and light effects on germination of Papaver bracteatum Lindl., P. orientale L., and P. somniferum L. Planta Medica 34(2): 135-143.
Bartlett, C.E.C., 1979. Effect of gibberellic acid and kinetin on tall bearded iris seed germination. The Iris Year Book, p. 89-93.
Barton, I.L., 1978. Temperature and its effect on the germination and initial growth of kauri (Agathis australis). New Zealand Journal of Forestry Science 8(3): 327-331.
Baskin, J.M. & C.C. Baskin, 1979. Promotion of germination of Stellaria media seeds by light from a green safe lamp. The New Phytologist 82(2): 381-383.
Bhiravamurty, P.V., Rama Das Kanakala, E. Venkata Rao & K.V. Sastry, 1979. Effect of some furofuranoid lignans on three species of seeds. Current Science 48(21): 949-950.
Biddington, N.L. & T.H. Thomas, 1979. Residual effects of high temperature pre-treatments on the germination of celery seeds (Apium graveolens). Physiologia Plantarum 47(4): 211-214.
Brencher, W., Luise Stange & K. Zimmer, 1978. Ersatz des Lichts bei der Keimung von Kakteensamen durch Gibberellinsäure. Gartenbauwissenschaft 43(2): 91-94.
Burger, S.J., J.O. Grunow & J.W. Rabie, 1979. Zaad- en ontkiemingsstudies op Antherphora pubescens Nees. Agroplantae 11(1): 15-17.
Burgmans, J.L., 1979. Pea seed treatment with methiocarb. Seed & Nursery Trader 77(5): 71.
Caminada, P., 1979. Germination of the pecan nut. Zimbabwe Rhodesia

Agricultural Journal 76(6): 237-238.
Campiranon, S. & W.L. Koukkari, 1977. Germination of wild rice, Zizania aquatica, seeds and the activity of alcohol dehydrogenase in young seedlings. Physiologia Plantarum 41(4): 293-297.
Carow, B., 1979. Samenkeimung von Gloriosa. Deutscher Gartenbau 33 (20): 845-846.
Carpita, N.C., C.W. Ross & M.W. Nabors, 1979. The influence of plant growth regulators on the growth of the embryonic axes of red- and far-red-treated lettuce seeds. Planta 145(5): 511-516.
Cerdonio, M., S. Morante-Mazzoncini & M. Vicentini-Missoni, 1979. Polar growth response of Pisum arvense L. seeds to weak magnetic fields. Canadian Journal of Plant Science 59(3): 883-885.
Chakrabarti, A.G., 1979. Effects of air pollutants on seed germinability. Bulletin of Environmental Contamination and Toxicology 21(6): 749-755.
Chaurasia, S.C. & D.P. Shaha, 1978. Effect of leaf extract of Pithecolobium dulci Syn. Mimosa dulci Benth on seed germination and seedlings growth wheat. Agriculture and Agro-Industries Journal 11(10): 9-11.
Chhipa, B.R. & P. Lal, 1978. Effect of pre-soaking of seeds with salt and hormone solutions and different quality water on wheat. Journal of the Indian Society of Soil Science 26(4): 390-396.
Chirco, E.M., J.R. Goodman & B.E. Clark, 1978. Comparison of 15-25 °C and 20 °C for the germination of flatpea, Lathyrus sylvestris. The Newsletter of the Association of Official Seed Analysts 52(3): 36-37.
Chirco, E.M., J.R. Goodman & B.E. Clark, 1979. Germination of deertongue (Panicum clandestinum L.) seeds. The Newsletter of the Association of Official Seed Analysts 53(3): 40-42.
Ciorlăus, A., 1977. The influence of certain herbicides upon the soybean seeds germination. Analele Institutului de Cercetari Pentru Cereale si Plante Tehnice - Fundulea, 42: 97-102.
Clausnitzer, I., 1979. Germinación de las semillas del dato o frutos del cardón, Lemaireocereus griseus (Haw) Britt & Rose. Revista de la Facultad de Agronomia, Maracaibo, Venezuela 5 (1): 351-365.
Clegg, M.D. & J.D. Eastin, 1978. A thermogradient generating sand table. Agronomy Journal 70(5): 881-883.
Collis-George, N. & M.D. Melville, 1978. Water absorption by swelling seeds. II. Surface condensation boundary condition. Australian Journal of Soil Research 16(3): 291-310.
Coumans, M., 1978. Rôle du pore basal dans la germination du glomérule de Betterave sucrière. Biologia Plantarum 20(2): 114-118.
Coumans, M., E. Ceulemans & T. Gaspar, 1979. Stabilized dormancy in sugarbeet fruits. III. Water sensitivity. Botanical Gazette 140 (4): 389-392.
Courtis, W.S., 1978. Rapid, contamination-free sowing of surface-sterilized seeds and spores. Canadian Journal of Botany 56(2): 225-227.
Cseresnyes, Z., 1977. A study of the qualitative indices of the sunflower seeds with a high oil-content, under less-than-optimum conditions. Analele Institutului de Cercetari Pentru Cereale si Plante Tehnice - Fundulea, 42: 103-109.
Currah, I.E., 1978. Fluid drilling. World Crops 30(1): 22-24.

Dadykin, V.P. & L.S. Kovadlo, 1979. Effect of vibration on seeds. Soviet Plant Physiology 26(2), part 2: 347-350.

Dafni, A. & M. Negbi, 1978. Variability in Prosopis farcta in Israel: seed germination as affected by temperature and salinity. Israel Journal of Botany 27(3-4): 147-159.

Damiano, C., G. Della Strada & R. Quarta, 1976. Studies on embrioculture technique to develop new early ripening peach scions. Annali dell'Instituto Sperimentale per la Frutticoltura, Rome, Italy, 7, 47-50.

Das, P.K., M. Kar & D. Mishra, 1978. Nickel nutrition of plants; I. Effect of nickel on some oxidase activities during rice (Oryza sativa L.) seed germination. International Journal of Plant Physiology 90(3): 225-233.

Devoy, J., 1979. Acid scarification. GC & HTJ 186(18): 21-23.

Duke, S.O., 1978a. Significance of fluence-response data in phytochrome-initiated seed germination. Photochemistry and Photobiology 28(3): 383-388.

Duke, S.O., 1978b. Interactions of seed water content with phytochrome-initiated germination of Rumex crispus (L.) seeds. Plant and Cell Physiology 19(6): 1043-1049.

Edmiston, J., 1979. The role of pigments in germination and dark-growth of white mustard (Sinapis alba L.) seeds. Biological Journal of the Linnean Society 12(3): 261-275.

Einhellig, F.A. & J.A. Rasmussen, 1978. Synergistic inhibitory effects of vanillic and p-hydroxybenzoic acids on radish and grain sorghum. Journal of Chemical Ecology 4(4): 425-436.

Ellis, R.H. & E.H. Roberts, 1979. Germination of stored cassava seed at constant and alternating temperatures. Annals of Botany 44(6): 677-684.

Emerson, B.N. & H.C. Minor, 1979. Response of soybeans to high temperature during germination. Crop Science 19(4): 553-556.

Esashi, Y. & Y. Tsukada, 1978. Thermoperiodism in cocklebur seed germination. Plant Physiology 61(3): 437-441.

Esashi, Y., Y. Tsukada & Y. Ohhara, 1978. Interrelation between low temperature and anaerobiosis in the induction of germination of cocklebur seed. Australian Journal of Plant Physiology 5(3): 337-345.

Evans, R.C. & D.G. Fratianne, 1977. Interactions of applied hormones in the germination of Lepidium virginicum seeds. The Ohio Journal of Science 77(5): 236-239.

Fawusi, M.O.A., 1978. Emergence and seedling growth of pepper as influenced by soil compaction, nutrient status and moisture regime. Scientia Horticulturae 9(4): 329-335.

Fawusi, M.O.A., 1979. Germination of Talinum triangulare L. seeds as affected by various chemical and physical treatments. Annals of Botany 44(5): 617-622.

Fowler, J.L., 1979. Laboratory and field response of preconditioned upland cottonseed to minimal germination temperature. Agronomy Journal 71(2): 223-228.

Gerson, R. & S. Honma, 1978. Emergence response of the pepper at low soil temperature. Euphytica 27(1): 151-156.

Gilbertson-Ferriss, T.L. & H.F. Wilkins, 1977. Factors influencing seed germination of Freesia refracta Klatt cv. Royal Mix. Horti-

cultural Science 12(6), section 1: 572-573.
Ginzo, H.D., 1978. Red and far-red inhibition of germination in Aristida murina Cav. Zeitschrift für Pflanzenphysiologie 90(4): 303-307.
Girton, R.E., 1979. Effects of oxygen concentration on the respiration of excised root-tip segments of maize and rice, and of germinating grains of rice and buckwheat. Physiologia Plantarum 46(1): 58-62.
Gomes, D.T. & A.E. Kretschmer jr., 1978. Effect of three temperature regimes on tropical legume seed germination. The Soil and Crop Science Society of Florida Proceedings 37, p. 61-63.
Gray, D., J.R.A. Steckel & J.A. Ward, 1979. The effects of a fluid sowing pregerminated seeds and transplanting on emergence, growth and yield of outdoor bush tomatoes. The Journal of Agricultural Science 93(1): 223-233.
Gunay, A.L. & P.S. Rao, 1978. In vitro plant regeneration from hypocotyl and cotyledon explants of red pepper (Capsicum). Plant Science Letters 11(3, 4): 365-372.
Gusta, L.V., K.J. Kirkland & H.M. Austenson, 1978. Effects of a brief magnetic exposure on cereal germination and seedling growth. Canadian Journal of Plant Science 58(1): 79-86.
Guta, M. et al., 1977. The influence of certain herbicides upon the quality of beans seed. Analele Institutului de Cercetari Pentru Cereale si Plante Tehnice - Fundulea 42: 239-243.
Guy, R., 1978a. Effets de la prégermination sur la germination de 14 espèces agricoles et potagères. Revue Suisse d'Agriculture 10(6): 185-188.
Guy, R., 1978b. La dormance des grains d'orge d'automme (variété Gold) et la stimulation de la germination par le Panogène. Revue Suisse d'Agriculture 10(4): 127-129.
Guy, R., 1979a. Nouvelles observations sur la maturité, la dormance, la germination et la levée des semences de fenouil. Revue Suisse de Viticulture Arboriculture Horticulture 11(5): 215-217.
Guy, R., 1979b. Observations sur la dormance et la germination des semences de fenouil. Revue Suisse d'Agriculture 11(3): 131-133.
Hageseth, G.T., 1978. Kinetic and thermodynamic parameters that describe isothermal seed germination. Journal of Experimental Botany 29(108): 281-293.
Hamilton, D.F. & P.L. Carpenter, 1977. Seed germination of Myrica pennsylvanicum L. HortScience 12(6), section 1: 565-566.
Harrison, C.R. & J. Arditti, 1978. Physiological changes during the germination of Cattleya aurantiaca (Orchidaceae). Botanical Gazette 139(2): 180-189.
Hasegawa, A., M. Goi, M. Sato & Y. Ihara, 1978. Fundamental studies on the asymbiotic seed germination of Calanthe. Technical Bulletin of Faculty of Agriculture, Kagawa University 29(2): 251-259.
Hassanyar, A.S. & A.M. Wilson, 1979. Tolerance of desiccation in germinating seeds of crested wheatgrass and Russian wildrye. Agronomy Journal 71(5): 783-786.
Haupt, W. & R. Scheuerlein, 1977. Short-term phytochrome response in seed germination. Zeitschrift für Pflanzenphysiologie 85(5): 445-450.
Hecker, M., 1978. Einfluss von Hydroxyharnstoff auf die DNA-Synthese

und Keimung der Samen von Agrostemma githago sowie Vaccaria pyramidata. Biologia Plantarum 20(3): 217-220.
Hegarty, T.W. & H.A. Ross, 1978. Differential sensitivity to moisture stress of seed germination and seedling radicle growth in calabrese (Brassica oleracea var. italica) and cress (Lepidium sativum). Annals of Botany 42(180): 1003-1005.
Hegarty, T.W. & H.A. Ross, 1979. Use of growth regulators to remove the differential sensitivity to moisture stress of seed germination and seedling growth in red clover (Trifolium pratense L.). Annals of Botany 43(5): 657-660.
Hendricks, S.B. & R.B. Taylorson, 1979. Dependence of thermal responses of seeds on membrane transitions. Proceedings of the National Academy of Sciences of the United States of America 76(2): 778-781.
Heslehurst, M.R., 1979. Germination of some Banksia species. Australian Plants 10(80): 176-177.
Heydecker, W., 1978. 'Primed' seeds for better crop establishment? Span 21(1): 12-14.
Hooper, G.R., R.P. Rice jr., S.L. Flegler & A.R. Putnam, 1978. The effect of ultrahighfrequence electromagnetic energy on adenosine triphosphate activity in germinating weed seeds. Journal of the American Society for Horticultural Science 103(2): 173-176.
Hopper, N.W., J.R. Overholt & J.R. Martin, 1979. Effect of cultivar, temperature and seed size on the germination and emergence of soya beans (Glycine max (L.) Merr.). Annals of Botany 44(3): 301-308.
Houck, J.R., 1979. A simple procedure for sowing orchid seeds. American Orchid Society Bulletin 48(10): 1019-1021.
Hundal, P.S. & H.N. Khajuria, 1979. Effect of GA_3 and thiourea on seed germination of different varieties of peach. The Indian Journal of Agricultural Sciences 49(6): 417-419.
Ichihashi, S., 1978. Studies on the media for orchid seed germination. II. The effects of anionic and cationic combinations relevant to seedling populations and culture periods on the growth of Bletilla striata seedlings. Journal of the Japanese Society for Horticultural Science 46(4): 521-529.
Iyer, C.P.A., R. Singh & M.D. Subramanyam, 1978. A simple method for rapid germination of pineapple seeds. Scientia Horticulturae 8(1): 39-41.
Jacobsen, J.V. & E. Pressman, 1979. A structural study of germination in celery (Apium graveolens L.) seed with emphasis on endosperm breakdown. Planta 144(3): 241-248.
Jacq, V. & P. Roger, 1977. Diminution des fontes de semis dues à la sulfato-réduction, par un prétraitement des graines de riz avec des Cyanophycées. Série Biologie Microbiologie du Sol Cahiers Orstom, Dakar, Senegal, 12(2): 101-107.
Jassem, B., M. Jassem & M. Deuter, 1978. Embryologische Ursachen der geringen Keimfähigkeit monokarpen tetraploiden Zuckerrübensaatgutes. Archiv für Züchtungsforschung 8(4): 263-269.
Jassey, Y., J. Monin, A. Cornu & P. Dommergues, 1979. Etude préliminaire des possibilités de germination chez les semences de Pétunia. Physiologie Végétale 17(1): 27:43.
Johnston, B.M. & G. Fernández H., 1979. Rol del pericarpio de Atri-

plex repanda en la germinaciòn. I. Efecto des lavado de los frutos en agua. Phyton 37(2): 145-151.

Julin-Tegelman, Å., 1979. The changes in endogenous cytokinin-like substances in Zea mays seeds during germination. Plant Science Letters 14(3): 259-262.

Kavi, P.S., 1977. The effect of magnetic treatment of soybean seed on its moisture absorbing capacity. Science and Culture 43(9): 405-406.

Kawahara, E., T. Wakamatsu & Y. Miura, 1977. Studies on a forage crop, Silphium perfoliatum L. II. On the germination of seed. Bulletin of Akita Prefectural College of Agriculture 52(3): 23-29.

Kawanabe, S., 1979. Relationship between germination temperature and the classification of grasses. Japanese Journal of Tropical Agriculture 23(4): 173-180.

Khare, L.J., 1978. Some factors affecting germination of seeds of Urgenia indica Kunth. I. Effect of light and water (soil moisture) on germination. The Journal of the Indian Botanical Society 57(4): 312-317.

Kiss, A.S., 1979. Germination stimulation by magnesium and the mechanism of action. Acta Agronomica 28(3-4): 384-390.

Kolbasina, E.I., 1978. Dynamics of gibberellin in vernalizing seeds of different winter wheat varieties. Physiology and Biochemistry of Cultivated Plants 10(1): 34-37.

Kordan, H.A., 1978. Effect of barbiturates on rice seedling germination and developmental behaviour. Annals of Botany 42(177): 73-81.

Kretschmer, M., 1978. Einfluss der Korngrösse bei Feldsalat (Vamerianella locusta L.) auf die Keimfähigkeit und Triebkraft. Gartenbauwissenschaft 43(5): 210-213.

Kumar, P. & Alka, 1978. Coumarin and gibberellic acid regulated seed germination and seedling growth of Raphanus sativus L. Comparative Physiology and Ecology 3(3): 175-177.

Labouriau, L.G., 1978. Seed germination as a thermobiological problem. Radiation and Environmental Biophysics 15(4): 345-366.

Labouriau, L.G. & A. Pacheco, 1978. On the frequency of isothermal germination in seeds of Dolichos biflorus L. Plant and Cell Physiology 19(3): 507-512.

Landgraff, A. & P. Junttila, 1979. Germination and dormancy of reed canarygrass seeds (Phalaris arundinacea). Physiologia Plantarum 45(1): 96-102.

Levert, J. & M. Lamont, 1979. Température et germination des glands de chêne pédonculé. Comptes Rendus des Séances de l'Académie d'Agriculture de France 65(12): 1006-1017.

Lewak, S. & A.A. Khan, 1977. Mode of action of gibberellic acid and light on lettuce seed germination. Plant Physiology 60(4): 575-577.

Lewis, J.A., G.C. Papavizas & N.R. O'Neill, 1979. Effect of seed immersion in organic solvents on germinability. The Journal of Agricultural Science 92(3): 563-570.

Lincoln, W.C. jr., 1978. Laboratory germination of Hieracium aurantiacum - Devils paint brush. The News Letter of the Association of Official Seed Analysts 52(1): 18-20.

Longo, G.P., M. Olginati, G. Rossi, M. Valente & C.P. Longo, 1978.

Effect of brief treatments with benzyladenine on growth and development of watermelon cotyledons. Plant, Cell and Environment 1(1): 39-43.

Lord, J.M., 1978. Evidence that a proliferation of the endoplasmic reticulum precedes the fromation of glyoxysomes and mitochondria in germinating castor bean. Journal of Experimental Botany 29(108): 13-23.

Lovato, A. & M.T. Amaducci, 1970. Problemi di analisi del seme di barbabietola (Problems with beet seed analysis). Sementi Elette 16(1-2): 33-41.

Loy, J.B. & K.B. Evensen, 1979. Phytochrome regulation of seed germination in a dwarf strain of watermelon. Journal of the American Society for Horticultural Science 104(4): 496-499.

Maftoun, M. & A.R. Sepaskhah, 1978. Effects of temperature and osmotic potential on germination of sunflower and safflower and on hormone-treated sunflower seeds. Canadian Journal of Plant Science 58(2): 295-301.

Mali, C.V., S.B. Varade, V.G. Musande & P.B. Chalwade, 1978. Critical soil water potential and seed hydration for germination of grain sorghum. Current Science 47(16): 587-588.

Mali, C.V., S.B. Varade & V.G. Musande, 1979. Water absorption by germinating seeds of sorghum varieties at different moisture potentials. The Indian Journal of Agricultural Sciences 49(1): 22-25.

Mangano, G., 1977. Influenza degli estratti acquosi di mais e di sorgo sulla germinabilità ed energia germinativa dei semi e sull' accrescimento delle plantule di specie diverse. L'Agricoltura Italiana 5/6: 391-409.

Manokaran, N., 1979. Germination of Malaysian palms. The Malaysian Forester 42(1): 50-51.

Mathiasen, R., 1978. Seed germination and infection techniques for three southwestern dwarf mistletoes. Plant Disease Reporter 62(1): 64-67.

McAdam, J.H. & P. Hayes, 1978. The effect of temperature on the germination of seven cultivars of Zea mays L. Record of Agricultural Research 26: 55-61.

Mehta, P.C., S.S. Puntamkar & S.P. Seth, 1979. Effect of pre-soaking of seeds in different salt with varying concentration on the germination and yield of wheat grown on salinised soil. Journal of the Indian Society of Soil Science 27(1): 99-100.

Mieseler, T.M., 1979. Seed geraniums. The Plant Propagator 25(1): 14-15.

Minton, E.B., D.F. Wanjura & J.D. Bilbro, 1979. Effects of premature defoliation and plant kill on germination of cottonseed. Agronomy Journal 71(4): 659-661.

Misic, P., N. Lekic & R. Todorovic, 1978. Improvement of seed germination in the grape vine. Savremena Poljoprivreda 26(11-12): 77-80.

Mohan Ram, H.Y. & R. Mazumdar, 1977. Effect of chlorflurenol on growth and geotropism of Cicer arietinum. Phytomorphology 27(2): 198-206.

Mohnot, K. & U.N. Chatterji, 1978. Thermo-physiological investigation on the imbibition germination of seeds of certain arid zone plants. II. The seeds of Prosopis cineraria. Transactions of Indian Society of Desert Technology and University Centre of Desert Studies 3(1): 35-39.

Momonoki, Y., T. Hasegawa, Y. Ota, Y. Kaneki & T. Suzuki, 1978. Studies on the germination of seeds of Bupleurum falcatum L. IV. The germination inhibitors of Bupleurum falcatum seeds. Japanese Journal of Crop Science 47(2): 197-205.

Muick, F., 1978. Propagation of Cypripedium species from seeds. American Orchid Society Bulletin 47(4): 306-308.

Mukherji, S. & A.K. Biswas, 1979. Modulation of chlorophyll, carotene and xanthophyll formation by penicillin, benzyladenine and embryonic axis in mung bean (Phaseolus aureus L.). Annals of Botany 43(2): 225-229.

Mullick, P.C. & U.N. Chatterji, 1978. Inhibition of seed germination and early growth of lettuce by joint action of vitamin K and IAA. Science & Culture 44(4): 177-178.

Muntean, L., 1977-78. Contributions a l'étude de la germination de graines du lupin blanc (Lupinus albus L.). Index seminum notulae botanicae Institutum Agronomicum 'Dr. Petru Groza' Hortus Agrobotanicus, p. 85-91.

Myers, E.G. & L. Carloni, 1978. Effect of various soil salinity regimes on germination and growth of fodder plants. L'Agricoltura Italiana 4/5: 285-299.

Nagao, M.A. & W.S. Sakai, 1979. Effect of growth regulators on seed germination of Archontophoenix alexandrae. HortScience 14(2): 182-183.

Nagy, J., K. Pasztor & J. Lazanyi, 1978. A study on the stimulation of seeds by ultrasonic treatment in maize. Növénytermelés 27(2): 141-150.

Nakayama, N., N. Iwatsuki & T. Asahi, 1978. Degenerative changes in properties of the mitochondrial inner membrane in pea cotyledons during germination. Plant and Cell Physiology 19(1): 51-60.

Narkiewicz-Jodko, M., 1978. Investigations of health state of seed material stored in silo bins without air access. Zeszyty Problemowe Postpów Nauk Rolniczych 202: 131-154.

Nayyar, V.L. & H.Y. Mohan Ram, 1977. Ultrastructure of starch grain breakdown in cotyledons of Cajanus cajan during germination. Phytomorphology 27(2): 187-190.

Nelson, S.O., W.R. Kehr, L.E. Stetson, R.B. Stone & J.C. Webb, 1977. Alfalfa seed germination response to electrical treatments. Crop Science 17(6): 863-866.

Ng, T.T. & T.H. Wong, 1977. Germination and seedling emergence of the tropical grass (Ischaemum magnum Rendle). The Malaysian Agricultural Journal 51(1): 7-14.

Niimi, Y., 1978. Influence of low and high temperatures on the initiation and the development of a bulb primordium in isolated tulip embryos. Scientia Horticulturae 9(1): 61-69.

Nikolaeva, M.G. & B.B. Yankelevich, 1979. Effect of low temperature on activity of some enzymes and on germination of yellow acacia seeds. Doklady Botanical Sciences 244-246: 6-7.

Nishimura, S. & E. Maeda, 1977. Histological studies of callus induction in rice seed. Japanese Journal of Crop Science 46(4): 275-285.

Nishiyama, I., 1978. Further evicence for the break on the Arrhenius plot of germination activity in rice seeds. Japanese Journal of Crop Science 47(4): 557-562.

Noronha, A., M. Vicente & G.M. Felippe, 1978. Photocontrol of germination of Cucumis anguria L. Biologia Plantarum 20(4): 281-286.

O'Brien, L. & R.A. Orth, 1978. An unexpected cause of poor germination and emergence of wheat. Crop Science 18(3): 510-512.

Okada, T., 1978. Studies on green panic seed. I. The use of seed volume weight in estimating quality of green panic seed. II. Method for measuring the volume weight of a small amount of seed. Journal of Japanese Society of Grassland Science 24(2): 133-141.

Okusanya, O.T., 1978. The effects of light and temperature on germination and growth of Luffa aegyptiaca. Physiologia Plantarum 44(4): 429-433.

Olivieri, A.M. & S.K. Jain, 1978. Effects of temperature and light variations on seed germination in sunflower (Helianthus) species. Weed Science 26(3): 277-280.

Padilla, C., J. Sarroca & G. Febles, 1978. Effects of sowing depth and humidity in two Cuban soils on the germination of Panicum maximum Jacq. Cuban Journal of Agricultural Science 12(3): 317-323.

Padmaja Rao, S., S. Appa Rao & R.K. Mukherjee, 1978. Germination and seedling growth of black gram as affected by seed size, colour and maturity. Indian Journal of Agricultural Research 12(3): 191-194.

Pahwa, S.K., S. Kumar, P. Bhardwaj & J. Prakash, 1979. Effect of prometryne under non-saline and saline conditions on early seedling growth of.Pennisetum typhoides Staff. and Hubb. Geobios 6(3): 111-113.

Pana, P., 1977. The effects of corn seeds cracking upon their seminal indices. Analele Institutului de Cercetari Pentru Cereale si Plante Tehnice - Fundulea 42: 89-95.

Pandey, B.N. & R.P. Sinha, 1978. Temperature effects on seed germination of Crotalaria spp. Geobios 5(4): 148-151.

Panov, N.V., N.P. Trunina & R.F. Trunin, 1977. Effect of pulse pressures on seeds of cereal plants. Biology Bulletin of the Academy of Sciences of the USSR 4(3): 360-362.

Pargney, J.C. & P. Thalouarn, 1978. Etude cytologique de l'embryon du Pin d'Alep lors de la germination: influence d'un traitement stimulant par administration successive d'ions mercure et chlore. Canadian Journal of Botany 56(23): 2931-2936.

Park, G.J., J.S. Yang, H. Han & J.Y. Lee, 1977. Studies to improve the germination of Dystaenia takesimana seeds. I. Influence of temperature and seed-wing removal on the germination of Dystaenia takesimana seeds. II. Influence of H_2O_2 and growth regulators on germination of Dystaenia takesimana seeds. The Research Report of the Office of Rural Development vol. 19 (Livestock), p. 99-103, 105-108.

Pathak, P.S., C.M. Misra & R. Deb Roy, 1978. Studies on the seeds and their germination in Robinia pseudoacacia Linn. Annals of Arid Zone 17(4): 363-369.

Pavlista, A.D. & J.G. Valdovinos, 1978. Changes in the surface appearance of the endosperm during lettuce achene germination.

Botanical Gazette 139(2): 171-179.
Pecket, R.C. & F. Al-Charchafchi, 1979. The photocontrol of respiration in light-sensitive lettuce seeds. Journal of Experimental Botany 30(117): 839-842.
Perino, C. & D. Côme, 1979. Conditions de germination de l'embryon de pommier et croissance de la racine et de l'hypocotyle de la plantule. Physiologie Vegetale 17(4): 829-838.
Perl, M., 1979. Invigoration of cotton seedlings by treatment of seeds for pregermination activities. Journal of Experimental Botany 30(114): 183-192.
Phaneendranath, B.R. & C.R. Funk, 1978. Germination stimulation of Kentucky bluegrass seed permeated with plant-growth regulators dissolved in acetone. Crop Science 18(6): 1037-1039.
Piccolo, A.L.G. & L.I. Thomazini, 1977. Nota sobre a germinação da semente de angico-do-cerrado (Piptadenia falcata Benth). Revista de Agricultura 52(4): 219-226.
Pingle, A.R. & V.R. Dnyansagar, 1979. Induction of germination in Solanum. Current Science 48(10): 449-450.
Pittman, U.J., J.M. Carefoot & D.P. Ormrod, 1979. Effect of magnetic seed treatment on amylolytic activity of quiescent and germinating barley and wheat seeds. Canadian Journal of Plant Science 59(4): 1007-1011.
Platonova, R.N., V.P. Ol'khovenko, G.P. Parfenov, A.A. Lukin & V.G. Chuchkin, 1977a. Effect of space flight factors and increased temperature on seeds of diploid and tetraploid buckwheat. Biology Bulletin 4(1): 47-53.
Platonova, R.N., G.P. Parfenov, V.P. Ol'khovenko, N.I. Karpova & M.E. Pichugov, 1977b. Germination of pine seeds in weight lessness on board the 'kosmos-782' satellite. Biology Bulletin 4(5): 628-634.
Porter, N.G. & S.M. Clark, 1979. Effect of temperature and light on the germination of seed of Solanum aviculare and S. laciniatum. New Zealand Journal of Experimental Agriculture 7(3): 307-310.
Porter, N.G. & H.M. Gilmore, 1976. Germination studies of the seed of Solanum laciniatum Ait. and S. aviculare Forst. New Zealand Journal of Experimental Agriculture 4(3): 343-345.
Potts, H.C., J. Duangpatra, W.G. Hairston & J.C. Delouche, 1978. Some influences of hardseededness on soybean seed quality. Crop Science 18(2): 221-224.
Pratt, L.H., 1979. Phytochrome: function and properties. Photochemical and Photobiological Reviews 4: 59-124.
Pressman, E., M. Negbi, M. Sachs & J.V. Jacobsen, 1977. Varietal differences in light requirements for germination of celery (Apium graveolens L.) seeds and the effects of thermal and solute stress. Australian Journal of Plant Physiology 4(5): 821-831.
Purohit, S.S., 1978. Effects of 2,4-dinitrophenol and ultraviolet radiation on germination and seedling growth of Helianthus annuus L. Comparative Physiology and Ecology 3(2): 89-90.
Rajpurohit, K.S. & D.N. Sen, 1977. Soil salinity and seed germination under water stress. Transactions of Indian Society of Desert Technology and University Centre of Desert Studies 2(1): 106-110.
Rao, D.V.M., J.S. Prakasa & S. Mahbood Ali, 1978. Studies on the effect of presoaking and media on the germination of hybrid

(Varalaxmi) cotton seed. The Mysore Journal of Agricultural Sciences 12(2): 223-225.
Ratajczak, L., W. Ratajczak & A. Hoffmannowa, 1979. The effect of hormone treatment of the seeds on growth and development of some Leguminosae. I. Hastening of germination and lupine, boradbean and pea seedling development by gibberellic acid. II. Genus differences in chlorophyll content changes under the effect of gibberellic acid. Bulletin de la Société des Amis des Sciences et des Lettres de Poznan. Série S Science Biologiques 18, p. 27-38.
Raza, S.H., 1977. Effect of temperature pretreatment on germination of seeds of Pennisetum typhoides var. HB 1. Indian Journal of Agricultural Research 11(4): 241-242.
Raza, S.H. & P.K. Saiprakash, 1979. Effect of 2,4-D seed germination hypocotyl elongation and amylase activity in Phaseolus radiatus. Current Science 47(23): 910-911.
Raza, S.H. & Y.D. Tiagi, 1977. Differential effect of various nitrogen sources on the germination of wheat. Geobios 4(5): 199-201.
Reid, J.S.G. & J.D. Bewley, 1979. A dual rôle for the endosperm and its galactomannan reserves in the germinative physiology of fenugreek (Trigonella foenum-graecum L.), an endospermic leguminous seed. Planta 147(2): 145-150.
Richter, E., 1978. Merkmalsdifferenzen zwischen Pal- und Markerbsen. III. Wasseraufnahme reifer Samen und Auflauffähigkeit bei konstant niedriger Temperatur. Gartenbauwissenschaft 43(1): 14-20.
Ritzel, R.E., 1977. Seed germination of Arabidopsis thaliana and its dependence on light and inhibitor substance. Arabidopsis Information Service 14: 40-46.
Rizk, T.Y., A.M. Al-Hasan, R.A. El-Tekriti & B.J. Alawi, 1978. Effect of salinity on germination and seedling vigor of some annual medics (Medicago spp.). Mesopotamia Journal of Agriculture 13(2): 105-121.
Roberts, H.A. & P.M. Lockett, 1977. Temperature requirements for germination of dry-stored, cold-stored and buried seeds of Solanum dulcamara L. The New Phytologist 79(3): 505-510.
Ross, H.A. & T.W. Hegarty, 1979. Sensitivity of seed germination and seedling radicle growth to moisture stress in some vegetable crop species. Annals of Botany 43(2): 241-243.
Ruge, U., H. Brunner & D. The, 1977. Ein keimhemmendes Prinzip der Futtermalve. Angewandte Botanik 51(3/4): 149-156.
Russo, G., 1977a. Observations of some biological characteristics of Clementine seeds. Annali della Facoltà di Agraria 29: 131-137.
Russo, G., 1977b. Germinability of Clementine seeds in relation to their number per segment. Annali della Facoltà di Agraria 29: 467-471.
Rytko, G., 1977. Study of laboratory of estimating seed-value of common orchard grass (Dactylis glomerata L.). Part I: Effect of biometric and aerodynamic traits on seed germinating ability. Hodowla Róslin Aklimatyzacja Inasiennictwo 21(6): 575-593.
Sahai, R., G. Kaur & P.S. Roy, 1977. Effect of some growth regulators and chemicals on seed germination and early seedling growth of Lathyrus aphaca Linn. Indian Journal of Ecology 4(1): 55-59.
Salazar, C.R. & A. Perez, 1976. Effect of three growth regulators on the germination of Carica papaya seeds. Revista ICA (Instituto

Colombiano Agropecuario) 11(3): 221-230.
Salehuzzaman, M. & M.K. Pasha, 1979. Effects of high and low temperatures on the germination of seeds of flax and sesame. The Indian Journal of Agricultural Sciences 49(4): 260-261.
Sanai, S., M. Nakayama & Y. Ota, 1977. Plant growth-regulating activities of nicotinamide. I. Effect of nicotinamide on growth of rice seedlings. Japanese Journal of Crop Science 46(1): 1-7.
Saraswathamma, D.N., 1979. The effect of pre-sowing hardening and foliar application with growth regulators in orach (Atriplex hortensis). Current Science 48(8): 357-359.
Satoh, S. & Y. Esashi, 1978. Rhythmic behavior in germination of cocklebur seeds. Physiologia Plantarum 43(3): 271-273.
Schopfer, P., D. Bajracharya & C. Plachy, 1979. Control of seed germination by abscisic acid. I. Time course of action in Sinapis alba L. Plant Physiology 64(5): 822-827.
Sett, R. & P.S. Ganapathy, 1977. Control of peg development in Luffa cylindrica. Phytomorphology 27(4): 383-389.
Sharma, B.K. & G.S. Lavania, 1978. Effect of some hormones on seed germination and early seedling growth of Vicia spp. Geobios 5(4): 176-178.
Sharma, B.K. & G.S. Lavania, 1979. Seed germination of Vicia hirsuta Gray and Vicia sativa L. Geobios 6(3): 133-135.
Sharma, V., 1978. Studies in germination of seeds in Welwitschia mirabilis Hooker. Acta Botanica Indica 6(1): 111-113.
Sheikh, A.S. & L.P. Mall, 1978. Water requirements of germinating cotton seeds. The Journal of the Indian Botanical Society 57(4): 317:321.
Sheoran, I.S., S. Babber & M.I. Khan, 1979. Water stress and starch accumulation in germinating guar (Cyamopsis tetragonoloba L.). Plant Science Letters 15(2): 159-163.
Shimizu, N. & N. Mochizuki, 1978. Studies on dormancy and germination of seed in finger millet (Eleusine coracana). II. Varietal differences of degree of dormancy and germination behaviour. Bulletin of the National Grassland Research Institute (Sōchi Shikenjō Kenkyū Hōkoku) 12: 76-91.
Shimizu, N., K. Tajima & R. Ogata, 1977. Studies on promotion of germination at low temperature in tropical grass seed. Bulletin of the National Grassland Research Institute 11: 47-56.
Singh, P.V., M.B. Singh & A.N. Khanna, 1973. Note on the temperature requirements of germinating rice seeds under controlled conditions. Indian Journal of Agricultural Sciences 43(4): 426-427.
Singh, B.D., R.B. Singh, R.M. Singh, Y. Singh & J. Singh, 1979. Effect of insecticides on germination, early growth and cytogenetic behavior of barley (Hordeum vulgare). Environmental and Experimental Botany 19(3): 127-132.
Smart, M.G. & T.P. O'Brien, 1979. Observations on the scutellum. I. Overall development during germination in four grasses. II. Histochemistry and autofluorescence of the cell wall in mature grain and during germination of wheat, barley, oats and ryegrass. Australian Journal of Botany 27(4): 391-411.
Smith, G.S., 1979. A note on the presence of water-soluble germination inhibitors in the seed pod of sainfoin (Onobrychis viciifolia Scop.). New Zealand Journal of Experimental Agriculture

7(4): 365-367.
Smith, A.R. & J. Van Staden, 1978. Changes in endogenous cytokinin levels in kernels of Zea mays L. during imbibition and germination. Journal of Experimental Botany 29(112): 1067-1075.
Spencer, N.E. & S.M. Siegel, 1978. Effects of sulfur and selenium oxyanions on Hg-toxicity in turnip seed germination. Water, Air and Soil Pollution 9(4): 423-427.
Srivastava, A.K. & N.K. Kooner, 1978. Effect of temperature on the physiology and biochemistry of germinating seeds of Phaseolus aureus Roxb. Acta Agronomica Academiae Scientiarum Hungaricae 27(3-4): 450-455.
Suciu, T., A. Salontai, L. Muntean, V. Felecan & L. Vaida, 1977-78. Recherches concernant la germination des semences de houblon (Humulus lupulus L.). Index seminum notulae botanicae Institutum Agronomicum 'Dr. Petru Groza'. Hortus Agrobotanicus, p. 79-83.
Takahashi, H., 1978. Seed germination ecology and cultivation of barnyardgrass after Italian ryegrass in unflooded paddy fields. Japanese Agricultural Research Quarterly (JARQ) 12(1): 44-48.
Takahashi, N., 1979. A concept of dormancy, germination and longevity in cereal seeds. The Reports of the Institute for Agricultural Research Tohoku University 30, p. 21-27.
Takeba, G. & S. Matsubara, 1977. Rapid disappearance of small fat bodies during the early stage of imbibition of lettuce seeds. Plant and Cell Physiology 18(5): 1067-1075.
Takeba, G. & S. Matsubara, 1979. Measurement of growth potential of the embryo in New York lettuce seed under various combinations of temperature, red light and hormones. Plant and Cell Physiology 20(1): 51-61.
Tamberg, T., 1979. The embryo cutting method for seed germination. The Iris Year Book, p. 93-95.
Tao, K.-L. & A.A. Khan, 1979. Changes in the strength of lettuce endosperm during germination. Plant Physiology 63(1): 126-128.
Thill, D.C., R.D. Schirman & A.P. Appleby, 1979a. The influence of soil moisture, temperature, and compaction on the germination and emergence of Bromus tectorum. Proceedings of The Western Society of Weed Science, vol. 32, p. 97.
Thill, D.C., R.D. Schirman & A.P. Appleby, 1979b. Osmotic stability of mannitol and polyethylene glycol 20,000 solutions used as seed germination media. Agronomy Journal 71(1): 105-108.
Thomas, J.F. & C.D. Raper jr., 1979. Germinability of tabacco seed as affected by culture of the mother plant. Agronomy Journal 71(4): 694-695.
Thomas, T.H., A.A. Khan & D.F. O'Toole, 1978. The location of cytokinins and gibberellins in wheat seeds. Physiologia Plantarum 42(1): 61-66.
Thomas, T.H., N.L. Biddington & D.F. O'Toole, 1979. Relationship between position on the parent plant and dormancy characteristics of seeds of three cultivars of celery (Apium graveolens). Physiologia Plantarum 45(4): 492-496.
Thompson, P.A., 1979. Some like it hot! GC & HTJ 185(4): 33-35.
Tinker, P.B., L. Reed, C. Legg & S. Hojer-Pederson, 1977. The effects of chloride in fertiliser salts on crop seed germination. Journal of the Science of Food and Agriculture 28(12): 1045-1051.

Tothill, J.C., 1977. Seed germination studie
tortus. Australian Journal of Ecology 2(4)
Townsend, C.E., 1977. Germination of polycross
vetch as affected by year of production. Crop
909-912.
Tran, V.N., 1979. Effects of microwave energy on th
seed coat and germination of Acacia seeds. Austra
Plant Physiology 6(3): 277-287.
Tsuzuki, E. & J. Watanabe, 1978. Studies on allelopathy
plants. Bulletin of the Faculty of Agriculture Miyazak
sity 25(2): 403-410.
Turn, N.S., N.V. Vorob'ev & T.P. Zhurba, 1979. Effects of is
solutions of dextran, sodium chloride, and sodium sulfate
germination of seeds and respiration of embryos in rice. Sov
Plant Physiology 26(2): 360-363.
Varsney, K.A. & B.D. Baijal, 1978. Synergistic and antagonistic behaviour of some growth regulators on germination of seeds of Pennisetum pedicellatum. Comparative Physiology and Ecology 3(3): 178-180.
Vatsa, V.K. & R.A. Sharma, 1978. Flurenol induced changes in the germination, seedling growth and yield of Trigonella foenum-graecum L. Comparative Physiology and Ecology 3(4): 243-245.
Vazquez-Yanes, C., 1979. Comunicaciones. Notas sobre la ecofisiología de la germinación de Cecropia obtusifolia Bertol. Turrialba 29(2): 147-149.
Verma, H.S. & P. Singh, 1978. Note on effect of seed treatment with IAA, IBA and GA on growth and yield of barley (Hordeum vulgare L.). Indian Journal of Agricultural Research 12(1): 59-60.
Vintilă, R., G. Lazăr-Keul, M. Keul & V. Soran, 1979. Phytotoxische Thirahexalin-Wirkungen auf Weizenpflanzen. Revue Roumaine de Biologie Série de Biologie Végétale 24(2): 133-141.
Visalakshy, A. & M.R.G.K. Nair, 1978. On the effect of phorate on germination of cowpea seeds and phytotoxicity to the seedlings. Agricultural Research Journal of Kerala 16(1): 102-103.
Watanabe, N., D. Tsuda & S. Hosokowa, 1979. Genetic studies on the dark-germination in tobacco seeds. II. Difference in genetic system between dark-germination and the sensitivity to the promotive red light effect in germination. Memoirs of the Faculty of Agriculture, Hokkaido University 11(3): 203-209.
Watt, L.A., 1978. Some characteristics of the germination of Queensland blue grass on cracking black earths. Australian Journal of Agricultural Research 29(6): 1147-1155.
Willatt, S.T. & R.G. Struss, 1979. Germination and early growth of plants studied using neutron radiography. Annals of Botany 43(4): 415-422.
Wilson, M.F. & E.A. Bell, 1978. The determination of the changes in the free amino acid content of the eluate from germinating seeds of Glycine wightii (L.) and its effect on the growth of lettuce fruits. Journal of Experimental Botany 29(112): 1243-1247.
Woolley, J.T. & E.W. Stoller, 1978. Light penetration and light-induced seed germination in soil. Plant Physiology 61(4): 597-600.
Yahiro, M., 1979. Effects of seed-pretreatments on the promotion of germination in papaya, Carica papaya L. Memoirs of the Faculty

ture Kagoshima University 15(24): 49-54.

, D.C. & D.E. Zimmer, 1978. Influence of harvest date and eezing on sunflower seed germination. Crop Science 18(3): 479-481.

Index*

* Plants, fungi etc. are indexed under their generic names.